Goldfish in the Parlour

Goldfish in the Parlour

The Victorian craze for marine life

John Simons

SYDNEY UNIVERSITY PRESS

First published by Sydney University Press
© John Simons 2023
© Sydney University Press 2023

Reproduction and communication for other purposes

Except as permitted under the Copyright Act 1968, no part of this edition may be reproduced, stored in a retrieval system, or communicated in any form or by any means without prior written permission. All requests for reproduction or communication should be made to Sydney University Press at the address below:

Sydney University Press
Gadigal Country
Fisher Library F03
University of Sydney NSW 2006
AUSTRALIA
sup.info@sydney.edu.au
sydneyuniversitypress.com.au

A catalogue record for this book is available from the National Library of Australia.

ISBN 9781743328729 paperback
ISBN 9781743328743 epub
ISBN 9781743328736 pdf

Cover image: Red rock cod (Scorpaena papillosa) and red handfish (Thymichthys politus). Chromolithograph after an illustration by Louisa Anne Meredith from her book *Tasmanian Friends and Foes, Feathered, Furred and Finned*, Marcus Ward, London, 1881.

Back cover image: "Goldfish" by Carlton Smith (1895).

Cover design by Miguel Yamin

Contents

List of Plates		vii
Preface		ix
1	The strange case of the missing fish	1
2	The Regent's Park Fish House and the Dublin Aquatic Vivarium	25
3	The domestic aquarium	71
4	She sells seashells	113
5	The public aquariums 1	159
6	The public aquariums 2	197
7	Australia: an imperial case study	231
Afterword: are the fish still missing?		265
Bibliography		273
Index		292

List of Plates

Plate 1 Photograph by Comte de Montízon (n.d.). Used with the gracious permission of Her Majesty Queen Elizabeth II and the Royal Collection Trust.

Plate 2 Photograph of Regent's Park Fish House exterior (n.d.) used with permission of the Zoological Society of London.

Plate 3 Photograph of Regent's Park Fish House interior (n.d.) used with permission of the Zoological Society of London.

Plate 4 Photograph of Dublin Aquatic Vivarium (n.d.) used with permission of the Zoological Society of Ireland.

Plate 5 "The Pets of the Aquarium" illustration sourced from George Kearley's book *Links in the Chain; or, Popular Chapters on the Curiosities of Animal Life* (1862).

Plate 6 "Goldfish" by Carlton Smith (1895).

Plate 7 "The Seaside in Autumn" by Robert Loudan Sr. (1858).

Plate 8 "A Contrast" by Abraham Solomon, *Illustrated London News* (1855).

Plate 9 "Seaside Sirens" by Jules Pelcoq, under the name of Jules Pelcor (1855).

Plate 10 Photograph by the author (n.d.).

Plate 11 "The Brighton Aquarium", *Illustrated London News* (10 August 1872).

Plate 12 "The Crystal Palace Aquarium" *Illustrated London News* (30 December 1871).

Plate 13 "The Aquarium at the Southport Pavilion" *Illustrated London News* (3 October 1874).

Plate 14 "The Yarmouth Aquarium and Winter Garden", *Illustrated London News* (1876).

Plate 15 "L'aquarium d'eau douce, dans le parc du Trocadéro" (1878).

Preface

I wrote this book in strange times and maybe you are reading it in even stranger times. While I was writing, the COVID-19 pandemic swept the world and continues to rage in most countries, fire and flood on an almost unprecedented scale visited the mainland of Australia and a war was raging in the Ukraine. Most of the work was done in Tasmania while we were largely virus free and lived lives which were very different from those lived almost anywhere else in the world. But the signs told us there was no room for complacency. A shark jumped up and pulled a child from a boat and two sharks attacked people in the inshore waters of our island where no sharks have been known before. Nearly five hundred whales were stranded on our west coast. Yellow-bellied sea snakes were seen swimming off our coast – much further south than their usual range. If I were writing this in sixteenth-century England, I'd see these events as monstrous portents and signs that the world is diseased.

This book may prove to be an elegy to a world that will soon be only dimly remembered. A friend in England recently sent me a quotation from Gilbert White in which he expressed his delight in being able to play with a more or less tame bat in his garden. Can we imagine a time when we won't fear catching a disease from bats?

It is not the first time I have written a book about animals during a health crisis. In 2002 I was writing *Animal Rights and the Politics of Literary Representation* in the midst of the outbreak of foot-and-mouth disease that led to the mass slaughter of farm animals in England. I remember driving through Kent with the reek from the pyres of burning bodies hanging over the fields in a cruel parody of the smoke which rises over the happy countryside in Virgil's *Georgics*, and the fumes from the disinfectant baths were sufficiently strong to fill the car. That passed.

The immediate occasion for this book was an invitation to give a paper at the Octopus Aesthetics Symposium organised by Dr Yvette Watt at the University of Tasmania in October 2018. This was part of her excellent Oktolab project. I put together a piece on the Crystal Palace octopus, which was one of those sensational animals that held the attention of the Victorian public for a time. In this case it was for a few weeks in 1871. I saw this paper (which was subsequently written up as a chapter titled "Cephallopodomania! Or, The Tentacular Spectacular" for Dr Watts' forthcoming book on octopus aesthetics) as a study of a prominent Victorian animal of the kind I had produced several times before. But as I explored the life of this octopus, I saw something that I had not noticed before. And that was that very little attention has been paid to fish in the field of animal studies or animal history. And I wondered why that was.

This led me to realise that being able to see a fish swimming in a glass tank, an experience we take for granted, was in the mid-nineteenth century something new and remarkable. People had simply not been able to see fish as they now could with the invention of the aquarium and all the things that went with it. So, in studying fish in Victorian England, I was also able to study a new set of experiences.

Like most of my work over the last 20 years, this book is about how people see and meet animals and, importantly, in what institutions and in what contexts these encounters happen. If you lived in, say, Harrogate in the 1860s, how would you have gone about seeing a fish?

As I try to answer that question, what emerges is a picture of the interconnection of ideas and technologies: the development of public zoos led to the development of public aquariums. The development of an interest in marine science as a domestic activity led to the craze for home aquariums. The need to exploit fish as an economic resource led to fish farms and marine science stations. The increase in leisure time towards the end of the 1840s led to the boom in class-segregated

Preface

angling – coarse fishing for the workers, game fishing for the gentry. People discovered the seaside both as a place to go on holiday and a place to learn about nature. This was all facilitated by developments in transport, science, engineering and finance. The combination of these things made it possible for a Victorian family from Harrogate to encounter fish in a public aquarium, or in the aquarium set up in their own parlour, or on the beach or on the banks of a river, canal or lake. The novel sight of a fish swimming in a glass tank was thus the highly determined product of a remarkable confluence of thoughts and things and may be considered as one of the most distinctive commodities of the Victorian period.

Another question that I explore is why there is so relatively little attention to fish in the animal rights literature. And I speculate briefly on what would a world look like where fish enjoyed the same welfare standards and protections against cruelty as other animals.

Finally, this book grew out of an older personal experience and one which I have spent much time considering over the years. In 2005 I visited Sydney Aquarium for the first time and there I encountered an enormous knob-headed wrasse. It swam to the corner of its tank and hovered at my eye level – I looked at it and it looked at me. I stood for what felt like a long time holding the fish's steadfast gaze and recognised that I was encountering a creature with what seemed to be some form of personhood and certainly the capability to choose what to do next. In 2009 or 2010 I visited the aquarium again. I went to the tank where the wrasse lived and it immediately swam up to me and hovered, meeting my eye as it had before. I recognised it at once. Did it recognise me? My pondering on that encounter was another route to this book. We are very ready to see personality of a kind in mammals and some birds but we are less often inclined to see individual traits in fish. The Victorian discovery of fish was not only an aspect of a scientific and leisure revolution; it was also the expansion of the relational field of humans. For the first time fish became our companions and a corner of many a Victorian parlour was given over to housing tiny fragments of their world enclosed in glass. Unfortunately, as I also show, this companionship came at a very high price for the fish.

I was much buoyed that just as the closure of Tasmanian borders was announced a whale went swimming down the channel outside my house entirely unimpressed by our rules. The picture on the cover of this book is a chromolithograph of a spotted handfish by the Anglo-Tasmanian naturalist and artist Louisa Anne Meredith. Spotted handfish are a very rare and threatened species that is found only in the section of the Derwent estuary directly opposite my house and the beach on which I walk my dog Teddy every day. I like to think that, while I was writing this book, these fascinating creatures were making their way around the seabed within a couple of hundred metres of my desk.

And, ironically, while this book was being written, London Zoo closed its Regent's Park aquarium, thus ending a history which began there with the opening of the first public aquarium in the world in 1853.

As with all works of this kind I have accrued many debts. I acknowledge the gracious permission of Her Majesty the late Queen Elizabeth II to reproduce the photograph of a pike by the Comte de Montízon in the Royal Collection. And I thank Karen Lawson, the Picture Library Manager of the Royal Collection Trust, for her friendly efficiency in helping me with the process. I thank Sarah Broadhurst of the Library of the Zoological Society of London for her help in securing a reproduction of the photographs of the Regent's Park Fish House, especially at a time when she had to negotiate the various stringencies and restrictions of the English lockdown. Christopher Schwitzer, Director of Dublin Zoo, kindly gave permission to use the photograph of the Aquatic Vivarium. Catherine de Courcy, whose books on the histories of Melbourne and Dublin zoos should be on every animal historian's shelf, provided not only a high-resolution image of the Aquatic Vivarium but also a copy of her index to the minutes of the Acclimatisation Society of Victoria (in its various manifestations) and pointed out a couple of entries in the minutes of the Royal Zoological Society of Ireland that I had missed. She also proved an informative and entertaining correspondent on many other zoo-related matters. My ex-student and regular correspondent Louise Ennis kindly gave me a copy of Yonge's *British Marine Life* which led me to details of deep-sea exploration that I had not picked up before. My dear friend of over 40 years and erstwhile colleague Professor Philip Martin told me what he would buy if he were a Victorian parson who had decided to take up fishing – I'd always thought a cricket stump and a stick of dynamite

Preface

would be quite sufficient for catching fish but he corrected me. My long-time friend and fellow old Balkan hand Dr Ruth Cherrington kindly checked the entries in Queen Victoria's journal for me – Благодаря ти много. My PhD student at the University of Tasmania, Michelle Murphy, sent me relevant details from the minute books of the Royal Society of Tasmania which she noticed as she was working on her own project. I also thank my former colleagues at Macquarie University, especially Professor Martina Mollering and Professor Bruce Dowton, for giving me a continuing academic home in Australia.

Lastly, my wife Kate, artist and silversmith, has taken an active interest in this project as she always does and helped me to see through the words to the pictures beneath.

Taroona, The Feast of Saint Sophronius of the Kiev Caves, 2022

A note on money, orthography and nomenclature

The Victorians used the pound sterling (£). This was divided into 20 shillings (s), each of which was divided into 12 pennies (d). A guinea was 21 shillings.

All weights and measures are given in the imperial quantities that the Victorians used.

Although aquarium has a Latin form it is, in my opinion, a modern coinage. I therefore use the English plural "aquariums" throughout rather than the pseudo-Latin "aquaria".

Octopus is a word with a Greek etymology. As a former Classicist I acknowledge the plural form "octopodes" which is good Greek and cringe at the commonly used "octopi" which is bad everything. However, I prefer the English form and use "octopuses" throughout.

The ichthyologist William Saville-Kent styled himself differently at different points of his life. I use his final formulation, William Saville-Kent, even when referring to periods when he himself used William Saville Kent or William Kent. To the class-honed English eye that hyphen makes all the difference.

1
The strange case of the missing fish

In 1873 the attendants at Brighton Aquarium noticed something very odd. Each day the tank that contained lumpfish seemed to have fewer fish in it. Yet there was no sign of any dead fish and the lumpfish were not known for eating each other. Nocturnal observation solved the mystery. When night fell the aquarium's octopus – presumably thinking it was unseen – clambered from its tank, climbed up into the tank occupied by the lumpfish (who must have dreaded the lights going out), ate a couple, then climbed out and slithered back to its own tank. Finally, it mistimed its adventure and was discovered, red tentacled:

> One morning, however, when Mr Lawler, one of the staff, on going to count our young friends, found an interloper amongst them. "Who put this octopus in No.27 tank?" he inquired of one of the keepers. "Octopus, sir? no one! Well, if he ain't bin and got over out of the next tank!" And this was just the fact.[1]

Alas, sometime later the octopus itself met a similar fate when it was put into another tank while its own was cleaned and was eaten by the incumbent nurse fish. But not before another outing when it and another octopus both escaped:

---

1   Lee, H., *The Octopus* (London: Chapman & Hall, 1875), p. 38.

One went west, the other went east; and, as if by preconcerted plan, neither was content merely to cross the frontier and visit his nearest neighbor, but both passed through, or over, one intervening tank and settled down amongst the tribes beyond.[2]

This time no harm was done as one of the escapees ended up in a tank of inedible crabs and the other found himself confronted by an aggressive lobster who forced him to hide behind some rocks.

This was by no means the first story of an octopus being where it shouldn't be. In the ancient world Aristotle had observed that octopuses could come out of the sea. Aelian recorded the troubling story of an octopus that would clamber out of the sewer via lavatory seats and go pilfering, while Pliny the Elder wrote of an octopus that climbed a tree on one of its extra-oceanic excursions.[3] So the classically educated readers of the aquarist Henry Lee's book *The Octopus* would have not been surprised to hear that an octopus was wandering around an aquarium bent on a nefandous purpose and would have been able to place it in a tradition of such stories validated by a prestigious ancient language. Although this appears to be the first case of an octopus leaving its tank in an aquarium, it wasn't the last. In April 2016, Inky, an octopus in the aquarium in Napier, New Zealand, had his 15 minutes of international fame when he climbed from his tank, squeezed into a sluice that fed directly into the adjacent ocean (obviously he had worked that out in some way) and escaped captivity.

Octopuses are very clever, and they also have many remarkable physiological features, not least their sensory system. Currently an enormous amount of research and thought is going into understanding their ways of experiencing and processing the world: their complex ability to feel and sense their environment; their curiosity and the appearance that they spend some of their time solving problems; their

---

2   Lee, H., *The Octopus* (London: Chapman & Hall, 1875), p. 41.
3   Aristotle, *The History of Animals*, Book IX, Part 37 (http://classics.mit.edu/ Aristotle/history_anim.9.ix.html); the references to octopuses in Aelian (*On the Characteristics of Animals*) and Pliny (*Natural History*) have been analysed by C.B.A. Ingemark, "The Octopus in the Sewers: An Ancient Legend Analogue", *Journal of Folklore Research*, 45 (2008), pp. 145–70.

1 The strange case of the missing fish

extraordinary life cycle and the sacrifice that female octopuses make to protect their young and ensure that some of them will survive; their complex social organisation; and their habit of collecting things and, it appears, using them for decoration (not unlike a bower bird).[4] It is by no means impossible that the two octopuses mentioned above did indeed somehow collude to explore their environment in different directions.

The missing fish from the lumpfish tanks or Inky missing from his tank led me to reflect on the fish that appear to be missing from the now enormous fields of animal studies and animal welfare. One can read thousands of pages of scholarly work yet hardly even come across a fish. Why is this? We understand fish better every day and a growing body of work on their physiology suggests that they are just as vulnerable to pain as mammals. And yet they still get little protection and images of piles of fish gasping for breath or just poured out of boxes onto quaysides cause none of the public anger and outcry that images of cruelty to retired racehorses in a Queensland abattoir can stimulate. An angler can happily pose with a half-dead fish in his or her hands while an unwise Facebook post of your holiday in South Africa paying to shoot lions perfectly legally will cost you your job. A sushi restaurant in Sydney can show its knife-wielding chef holding up a living fish that is clearly distressed but I doubt if customers would go into a steak house that showed the chef brandishing a bolt gun at the curly forelock of a cute little bullock. What is it about fish that seems to militate against the duty of care that we might normally feel towards animals and why is what protection they do have usually designed to preserve their stock numbers for future exploitation? Where did the phrase "I'm a vegetarian, I'll have the fish" come from?

The problem may start from the definition of an animal in Peter Singer's foundational text *Animal Liberation*. Here it is argued that the

---

4 I once attended a paper on the octopus colonies of Eagle Neck in Tasmania. The presenter, who was also a diver, showed extensive underwater film, including some sections which showed octopuses with decorated lairs and octopuses collecting things for this purpose. He said that this was previously unknown behaviour. On the way home I dropped into a local Indigenous art gallery where I saw a large print of an octopus from the Torres Strait Islands. The octopus was shown sitting on a highly decorated mound. This is the sort of thing that makes Indigenous people angry.

point at which we might start to consider our approach to the welfare and care of any creature was at a line drawn:

somewhere between a shrimp and an oyster.[5]

Subsequently there has been a good deal of argument over oysters, especially whether it is acceptable for them to be eaten as part of a vegan diet, and Peter Singer himself is now more tentative as to where this line might be drawn. But a line is drawn here at a point where consciousness and neural networks appear not to have developed in such a way that the object of our attention might feel pain (or know or care very much what we do to it). This argument from pain will be explored in a little more detail below but I think it is a curious one and no different in its form from arguments that refer to intelligence or emotional sensitivity as a reason for an enhanced welfare status (for example, it is fine to use mice for vivisection but not primates). What it lacks is an appeal to being. If an oyster cannot feel pain (and the presence or absence of recognisable physiological structures is no guarantee of that either way) then surely it still has an existence that commands our concern and respect.

Take, for example, the case of Ming the Mollusc. Ming was a clam born in 1499 or thereabout and was very possibly the oldest living thing, save lichens, on the planet. He was fished up by the learned scientists of Bangor University who had put out to sea to kill things as a way of finding out about climate change. They put him in the freezer where he died. Only then was it noticed how very old he was and what a remarkable creature. And yet the way this incident was reported was, generally, rather amused in tone. The name Ming the Mollusc isn't a good start. But can you imagine what would happen if a university scientific expedition killed a 200-year-old tortoise or a 100-year-old parrot or even an ancient whale. I think there would be outrage and outcry and bad publicity. Internal inquiries instituted at the university. A redesign of the animal ethics forms at the very least. But Ming just died before his time – and who knows what his time would have been – for no good reason. People appear to have found it amusing.

---

5   Singer, P., *Animal Liberation* (New York: Avon Books, 1975), p. 16.

1 The strange case of the missing fish

The case of Ming brings us back to Peter Singer (I think Ming would not have made the cut as an animal using Singer's definition which may be another reason his untimely death caused mirth rather than outrage) and the uncertainty about the status of fish (or sea creatures) from the very beginnings of the animal rights movement and thus in the discourses of animal studies. This uncertainty also fatally affects the claim of fish as moral patients deserving our consideration. We can understand how Singer's definition came about when we consider his uncompromising utilitarianism in a perspective structured by Jeremy Bentham's famous argument about animal suffering (made in the context of his call to extend the circle of legal and moral concern to all sentient beings):

> The question is not can they reason? Nor, can they talk? But can they suffer?[6]

The problem with this is that if we assume that fish (or oysters) cannot suffer – and that is a widely held view which has been scientifically challenged for some time now – then it follows that, in a utilitarian framework at any rate, we need not concern ourselves with their welfare and that any harms done to them are of a lesser order that those we might do to mammals. If, indeed, you can do them harm at all.

Interestingly, an exception to the problem of the missing fish may be found in the first public action of the American Society for the Prevention of Cruelty to Animals when its founder, the activist Henry Bergh, boarded a turtle ship and arrested the captain for animal cruelty. The case went to court, and it was successfully argued by the defence that as turtles were fish they could not suffer and that, therefore, the apparent cruelty meted out to them as alleged by Bergh could not have happened. Although this case failed to win on the day, two important points had been made. The first was that it was possible for an animal

---

6   The original quotation is from Jeremy Bentham's *An Introduction to the Principles of Morals and Legislation*. This is cited by almost everyone concerned with animal rights and animal welfare and, as I shall argue later in this book, is the source of many regrettable developments in thinking about animals.

welfare organisation to advocate for the welfare of a fish. Secondly, that it was possible to debate in a formal setting such as a court of law whether a turtle was a fish or not and what exactly was the difference between a fish and any other animal.[7]

I am not arguing, of course, that a reading of Peter Singer gives carte blanche to treat fish in any way we like. It is rather that the lack of a fully developed understanding of the physiology and neural networks by which fish might experience pain and other forms of discomfort, including anxiety and fear, has inhibited the growth of a discourse of concern and welfare for fish which might begin to put them on par with other animals so far as limited legal protection from deliberate cruelty is concerned. I therefore feel that the acceptance of at least this part of the Benthamite program by the general public – as most people would agree that it seems wrong to cause an animal pain for no reason (we may tolerate this in agriculture and medical research provided we don't see it and believe it is for our benefit) – has not extended itself to fish. It is also interesting that Singer chose to draw his line between two marine creatures. Surely there would have been the option to draw it between two terrestrial creatures: an earthworm and a maggot for example. While earthworms have neural structures that look as if they would suffer pain, maggots probably do not and the defensive reactions that they and some other insects exhibit when stressed or mutilated are often described as reflexes. But the point is not where the line might be drawn but rather that, when Singer came to draw it, his mind more naturally turned to sea creatures than to land creatures, and this surely tells us something about a hierarchy of concern and moral status which underpins a great deal of the standing of fish and the way that they are generally considered.[8] And this problem in ethics and animal welfare has inevitably lapped over into animal studies which are, by and large,

---

7   See Freeberg, E., *A Traitor to His Species* (New York: Basic Books, 2020), pp. 10–18 and Burnett, D. G., *Trying Leviathan* (Princeton, NJ: Princeton University Press, 2007) which looks at other early legal cases revolving around the status of marine mammals.
8   The question of pain in fish and other marine animals is addressed later in this book. The organisation Crustacean Compassion campaigns for recognition of the pain felt by crustaceans and for their legal protection. (https://crustaceancompassion.org.uk)

1 The strange case of the missing fish

philosophically grounded in thinking about animal rights and animal welfare.

In his recent and extremely valuable study of the history of Bristol Zoo, Andrew Flack makes the point very well:

> Until very recently, fish were habitually omitted from the category of creatures we term "wildlife"; their extensive use as sustenance, in particular, rendered them marginal entities in human ideas about wild things ... Though they were far from unpopular exhibits, fish clearly were positioned and perceived at the Zoo less as recognizable individuals and more as striking oddities emanating from mysterious aquatic depths, and thus they were well out of any humanist moral circle.[9]

There has been at least one RSPCA prosecution of a seafood seller for cruelty to lobsters. But at the same time, the prawn industry routinely cuts off the eye stalks of live female prawns to promote reproduction and prawns remain on the menu with no indication as to whether they have been produced using "eye-stalk ablation". In thinking about the prawns he has eaten and the prawns he has spent hours observing on a Scottish beach, Adam Nicolson finds himself forced to reflect on:

> the stupidity of my encounter with them.[10]

I am fairly certain (even in Australia where animal welfare laws appear to be rarely enforced) that if the same welfare standards and legal protections were applied to fish as are applied to mammals and birds, virtually every fishing boat landing its catch, every live-fish processing plant or market stall or fishmonger's shop, every fish farmer, every private angler, every restaurant serving seafood could be prosecuted.[11]

---

9 Flack, A., *The Wild Within* (Charlottesville: University of Virginia Press, 2018), p. 128.
10 Nicolson, A., *The Sea Is Not Made of Water* (London: HarperCollins, 2021), p. 42.
11 Linzey, A. and Linzey, C., *The Global Guide to Animal Protection* (Urbana: University of Illinois Press, 2013) is a comprehensive review of animal welfare laws.

This is quite a stunning notion when thought of with the framing question: "What would the world be like if fish were legally protected?" We should also note the general neglect of Darwin's work on the emotions in animals – which he thought was every bit as valuable and important as his work on evolutionary physiology. My guess is that this is because it is easier to justify vivisection if you simply ignore half of the work of the founding father of the science you practise by putting the suggestion that the animals you work on may have thoughts and feelings into a cone of silence. But if this applies to mammals how much more must it apply to fish?

Of course, it helps to be an animal that Westerners don't eat. Think of the outcry over the cruelly discarded racing greyhounds in New South Wales a few years ago compared with the continued toleration of eggs produced in batteries. It is also true that the closer an animal is to pet status, the higher its chances of attracting attention if its welfare is threatened, and, as we move along the spectrum of agriculture, these chances diminish to the extent that the animals are perceived to be useful for food or some other purpose until we reach fish, which have no chance at all. Wild animals also have a higher chance of attracting welfare concerns if they are cute or "iconic", so a Tasmanian devil's chance of having someone take up the legal cudgels on its behalf are somewhat better than a funnel web spider's. Recently an Australian politician spoke scathingly about those who opposed new dam building on the grounds that an endangered moth would be further threatened. Would he have felt so able to make this point if the animal in question had been a koala?

It doesn't take much imagination to understand how this situation might have come about nor how the hierarchies of concern and interest that appear to structure the way we live with other animals form themselves. So, when I started to think about fish, part of that thought was always going to remain, to some extent, outside the main currents of animal welfare and animal rights discourse. However, this book is only partly about the welfare issues that present themselves when we consider fish and marine life in general. It is mainly about the ways in which these things became visible in Victorian England and thence to the world at large in the second half of the nineteenth century and how this visibility stimulated an interest that from time to time became a

1 The strange case of the missing fish

full-scale international craze. This is the fourth of my books on animals in the Victorian period and my sixth on animals; each of them has had, to a greater or lesser extent, the program of exploring and promoting, by means of a historical or cultural analysis, the welfare of the animals about which I am writing, whether this is through the development of a taxonomy of representational tropes, looking at the contradictory values assigned to kangaroos in Australia or giving some agency to a long-dead hippopotamus.

Fish are largely missing from the growing accounts of animals in history. Perhaps the most obvious sign of this can be seen by perusing the list of animals featured in the excellent and seminal *Animal* series published by Reaktion Books which, for more than a decade now, has surfed the growing wave of interest in animals not only within animal studies but also in the academic and cultural world more generally and has played for animal studies the role that the *New Directions* series played for literary theory in the 1980s. At the time of writing (late 2020), 100 titles had appeared or had been announced of which only seven featured a fish. They are *Goldfish*, *Salmon*, *Trout*, *Octopus*, *Squid*, *Sardine* and *Shark*. There are also *Turtle*, *Lobster*, *Oyster*, *Dolphin* and *Whale*.[12] These animals are not technically fish but they will feature from time to time in this book (Victorians were not always 100 per cent sure about the fishy status of sea mammals) and shellfish (or at least their shells) play an important role in the Victorian encounter with sea life and will be explored in some detail later in the book. So only 12 of the 100 titles or 12 per cent of an extensive and impressive list feature animals who live in the sea and only 7, or 7 per cent, feature actual fish. This strikes me as a remarkably small proportion and certainly not reflective of the actual numbers of fish compared with mammals, birds

---

12   Boos, A.M., *Goldfish* (London: Reaktion Books, 2019); Coates, P., *Salmon* (London: Reaktion Books, 2006); Owen, J., *Trout* (London: Reaktion Books, 2012); Schweid, R., *Octopus* (London: Reaktion Books, 2014); Wallen, M., *Squid* (London: Reaktion Books, 2021); Day, T., *Sardine* (London: Reaktion Books, 2018); Crawford, D., *Shark* (London: Reaktion Books, 2008); King, R.J., *Lobster* (London: Reaktion Books, 2011); Stott, R., *Oyster* (London: Reaktion Books, 2004); Rauch, A., *Dolphin* (London: Reaktion Books, 2013); Roman, J., *Whale* (London: Reaktion Books, 2005).

and even insects (which feature in 9 books or 9 per cent of the number of titles and so are better represented than fish).

So where are the fish? At the most obvious and superficial level a publisher will, if they want to stay in business, produce books that will sell, and books called *Lion*, *Kangaroo*, *Cat* or *Elephant* are surely likely to sell better than books called *Gudgeon*, *Dace*, *Plaice* or *Flathead*. But why should this be so? I can see that many fish don't have a great deal of charisma but what about *Pike*, the master predator of the fresh waters of Europe, or *Stickleback*, with its heroic and colourful breeding cycle, or *Carp*, the staple food of medieval England and central Europe and still to be seen, grown to enormous size, in the moats of places like Bodiam Castle, or *Ray*, deadly and gliding like a stealth bomber, or *Barramundi*, as iconic in its way as the kangaroo or emu? These books might sell well, so an equally likely explanation is the absence of people sufficiently interested in fish to be prepared to write about them in this way. And to understand why that is we must go back to the reasons for the relative absence of fish in animal rights and animal welfare thinking sketched out above as the majority (but by no means all) of the Reaktion authors are people involved in some way or another in academic animal studies where they won't be used to meeting many fish.

In addition, although there has been a growth in other kinds of writing about specific animals – specific historical elephants, giraffes, hippos and rhinos, for example – I can only think of a handful of titles that address fish. Books on cod, oysters, lobsters and eels spring to mind.[13] So the relative scarcity of books about fish in the modern animal studies literature can't be reduced to one flagship series but appears to be an issue that is structural to the field of writing about animals (and thus reading and thinking about animals).

There has also been a considerable growth in books about zoos, circuses and other forms of animal display. This has not been matched by very much on fish or aquariums. There are some clever articles and a very recent book by Silvia Granata, an interesting article by Christian

---

13 Kurlansky, M., *Cod* (London: Vintage, 1999); Kurlansky, M., *The Big Oyster* (New York: Random House, 2007); Fort, T., *The Book of Eels* (London: HarperCollins, 2003); Corson, T., *The Secret Life of Lobsters* (New York: Harper Perennial, 2004).

1 The strange case of the missing fish

Reiss, Judith Hamera's *Parlour Ponds*, which is a very thoughtful study of the "cultural work" of American home aquariums up to the 1970s, and Berndt Brunner's *The Ocean at Home*, which is a valuable survey particularly strong on European aquariums but which is too brief to offer more than a taste of the immensely complex history that unfolds around the capture and display of live fish.[14] Then there is Mareike Vennen's truly excellent *Das Aquarium, Praktiken, Techniken und Medien der Wissenproduktion (1840–1910)* (*The Aquarium, Practices, Techniques and Media of Knowledge Production*) which is written in quite difficult academic German and will probably never be translated.[15] Also in German, Ursula Harter's *Aquaria im Kunst, Literatur und Wissenschaft* (*Aquaria in Art, Literature and Science*) looks not only at aspects of aquarium history but also at the cultural

---

14 Granata, S., "Let Us Hasten to the Beach: Victorian Tourism and Seaside Collecting", *Lit: Literature, Interpretation, Theory*, 27 (2016), pp. 91–110; Granata, S., "At Once Pet, Ornament and Subject for Dissection: The Unstable Status of Marine Animals in Victorian Aquaria", *Cahiers Victoriens et Édouardiens* (en ligne), 88 (2018); Granata, S., "The Dark Side of the Tank", in H. Kingstone and K. Lister, *Paraphernalia! Victorian Objects* (London: Routledge, 2018); Granata, S., *The Victorian Aquarium* (Manchester: Manchester University Press, 2021); Reiss, C., "Gateway, Instrument, Environment: The Aquarium as a Hybrid Space between Animal Fancying and Experimental Zoology", *International Journal of History and Ethics of Natural Sciences, Technology and Medicine*, 20 (2012), pp. 309–36; Hamera, J., *Parlor Ponds* (Ann Arbor: University of Michigan Press, 2011); Brunner, B., *The Ocean at Home* (London: Reaktion Books, 2011). Silvia Granata's book came out just as this one was finished and I did have time to see a copy. It deals largely with the home aquarium movement and takes a broadly literary approach through aquarium manuals and handbooks. The timing was fortunate as the cover illustration is the same as that originally intended for this book – a chromolithograph of a wrasse from Gosse's *The Aquarium*; fortunately, the Louise Meredith handfish came to mind and I think that this offers a much more appropriate cover for a Tasmanian book.
15 Vennen, M., *Das Aquarium* (Göttingen: Walstein Verlag, 2018). See also her "Echte Forscher und wahrer Liebhaber, der Blick ins Meer durch das Aquarium im 19 Jahrhundert" (Genuine researchers and true enthusiasts, the glimpse into the sea through the aquarium in the 19th century) in A. Kraus and M. Winkler (eds), *Weltmeere: Wissen und Wahrmehrung im Langen 19 Jahrhundert* (Worldseas: Knowledge and Perception in the long 19th century) (Göttingen: Vandenhoeck & Ruprecht, 2014), pp. 84–102.

legacy of aquariums up to fish-tank screensavers.[16] In addition, there is Rebecca Stott's short book *Theatres of Glass* which tells the (highly contested) history of the scientific discovery of the chemistry of sea water and its aeration, and of the first successful experiments in keeping sea creatures alive under conditions of domestic captivity.[17] Recently Adam Nicolson has reinvented the study of rockpools in the spirit of Philip Gosse (who will play a significant role later in this book).[18] In the USA, Susan Shetterly has written about the importance of seaweed both to the life of coastal communities and to the environment.[19] The history of marine and littoral science is, of course, not the same as the history of fish but this book moves between the two as between two distinct but interlaced narratives.

Why has more attention been paid to the institutions in which land animals are displayed than those in which sea animals are displayed? As I shall show, the boom in municipal aquarium building in Victorian England (and across Europe, the USA and Australia) more than matched the boom in municipal zoo building and the reasons for the failure or success of individual projects are very similar indeed to those we find when we look at Victorian zoos. The materials for the study of aquariums are marginally harder to come by but that wouldn't put off a determined PhD student or a more senior scholar with time on his or her hands and nor would the absence of any significant secondary literature. I also wonder if there is a gendered inflection here. Most of the important figures in the story of Victorian zookeeping are men whereas the world of Victorian marine sciences is significantly feminised. Marine science was thought of as more suitable for ladies than the sex- and reproduction-obsessed disciplines of botany and zoology – although, as Helen Ellis has shown, the nature of masculinity in the context of science was by no means unproblematic.[20] In addition,

---

16 Harter, U., *Aquaria im Kunst, Literatur und Wissenschaft* (Heidelberg: Kehrer Verlag, 2014).
17 Stott, R., *Theatres of Glass* (London: Short Books, 2003).
18 Nicolson, A., Nicolson, A., The Sea Is Not Made of Water (London: HarperCollins, 2021), p.42.
19 Shetterley, S.H., *Seaweed Chronicles* (New York: Algonquin Books, 2018).
20 Ellis, H., *Masculinity and Science in Britain 1831–1918* (London: Palgrave, 2017).

1 The strange case of the missing fish

the domestic routines of aquarium keeping and other sea-related pastimes are frequently associated with women and, as will be seen, this appears to be borne out by the popular iconography of the domestic aquarium. Some of the most important discoveries in the field were made by women and women played a prominent role in many other aspects of marine science. But, of course, most of these women were excluded from the major scientific institutions (at least until very nearly the end of our period) and, although some of them published work which was highly respected by their male colleagues – Sarah Bowdich Lee, for example, was greatly admired by Georges Cuvier and was his first biographer – they did not have the same platform as those male scientists.[21] So, writing the history of fish and marine science in Victorian England involves, in part, writing the history of a band of female scientists who, in the way of things, have still not got the recognition they deserve. As Barbara Gates has pointed out, the wonderful illustrations in Gosse's *The Aquarium*, which were one of the most influential forces on driving the craze, were the work of Elizabeth Bowes Gosse, Gosse's wife, but they are not signed by her; the plates are labelled P.H. Gosse, delt, [delineated] and she is nowhere mentioned in the book.[22]

If we were to try to define the history of fish and meetings with fish as a problematic, that is, as a field described by the questions one might ask about it, those questions would mainly address the absence of that history. Why is there such a small literature about fish? Why do fish not have the same legal status as other animals? Why do many people who eat fish see no problem with claiming to be vegetarian? Is the prominence of women in nineteenth-century marine science somehow related to the neglect of fish as objects of scholarly attention? Why are mammals more interesting than fish? This is a telling activity as an exercise like this would normally engage a much more positive vocabulary and an exploration of things that are present rather than things that are absent.

---

21 Orr, M., "Women Peers in the Scientific Realm: Sarah Bowdich (Lee)'s Expert Collaborations with Georges Cuvier, 1825–1833", *Notes and Records of the Royal Society*, 69 (2015), pp. 37–51.
22 Gates, B., *Kindred Nature* (Chicago: University of Chicago Press, 1998), p. 74.

It appears then that fish occupy what might be thought of as a negative discursive space. This work is an attempt to fill that space and to replace it with a set of more positive fields. To this end, one of the several organisational principles of this study is to consider the encounter with fish as happening in three spatial domains: the public institution, the private home and managed nature. The three spaces are sometimes separate and sometimes overlapping, and it is within one or more of these that the Victorian encounter with fish was produced and enacted.

The value of thinking about the issue in this way is that it focuses our attention on the structural determinations of a predominantly existential phenomenon. The ability to encounter a fish, to watch fish or to catch a fish is made possible only within a specific part of one of the three domains and the institutions they contain; without these domains we might not easily be able to recognise the uniqueness in a historical moment of that time when the Victorians discovered fish and began to explore the sea and its shores.

In addition, the very novelty of seeing a fish swimming in a tank was not just a new experience for the Victorian public. It was also a very specifically manufactured and determined product created by the complex intersection of a range of economic, technical and social developments. These include more money and more leisure time after the deep recessions of the 1830s and 1840s; the abolition of tax on glass in 1845 (very significant as it could amount to as much as 300 per cent of the value of the product); developments in plate glass technology and manufacture; the discovery of the seaside; the movement to manage and "improve" working-class leisure time through public institutions of culture and education; the development of accessible commercial models for the raising of venture capital through share issues after the passing of the revised *Joint Stock Companies Act* in 1856; an ever-extending railway network offering cheaper fares; steamships and, later, the Suez Canal, cutting journey times to the colonies and elsewhere; improvements in cargo handling and packaging that reduced the previously shocking levels of mortality in live transport; the scientific revolution led by Darwin and his followers but also by many others in different fields and the growing general interest in science, especially as a domestic pursuit; developments in paper manufacture

1 The strange case of the missing fish

and the printing press that made books cheaper and stimulated the growth of magazines and newspapers; shifts in the balance of population from rural to urban; Sabbatarian and creationist controversies; and the boom in municipal zoo building. And this is by no means an exhaustive list.

The interactions of causality enmeshed in this list are contextualised by an increasing sense that art and science exist in different domains. Although it might be possible to put a date on this, it is better to think of it as a mobile process which develops in the first half of the nineteenth century to the point that, by 1854, Dickens is able to use it as a key narrative device in *Hard Times*. This epistemological division had significant impact for animals and fish as it destabilised their place in theological, aesthetic and scientific discourses. At the same time, developments in colour printing enabled representations of fish which were richer and brighter than anything that had been seen before. And here a paradox emerges: just at the time when fish were increasingly the object of ichthyology and related marine sciences and were being developed as part of the economic value chain that would turn them into precious food commodities, their aesthetic representation positioned them as subjects as never before. So, the scientific world of an aquarist like Philip Gosse is balanced by the aesthetic world of the beautiful chromolithographs that illustrated his books. And in these and other representations of fish which were not mere diagrams we see an apprehension of the sensual and emotional world that they inhabit, which is positioned on a profound fault line in the Victorian intellectual landscape. The idea that a fish has a life like any other animal is well expressed by John Harper in his *Glimpses of Ocean Life* (which was revealingly subtitled *Rockpools and What They Teach Us*):

> Now whether we look at the singular skill of a bird building its nest, the hen sitting near and protecting its brood, or the cat grasping her young in its jaws and carrying them home to safety, we shall find that the charming traits are wonderfully combined in the humblest members of the finny tribe, viz., the common stickleback – the little creature that boys catch by the thousands

with a worm and pin, – that lives equally content in the clear blue
sea or the muddy fresh water pool.[23]

This makes it clear that what we learn from rockpools is not merely
a scientific understanding of the physiology and the life cycle of the
creatures that inhabit them but, more importantly perhaps, an
apprehension of the richness of their being, especially as this serves to
display the wonderful plan of divine creation and the important parts
that even the smallest and commonest animals have to play within
it. The case of cruelty to turtles brought by Bergh in New York was
mentioned above. Another case where a gentleman showed sensitivity
to the suffering of a marine creature, was the discovery that a porpoise
that Frank Buckland was attempting to keep alive for London Zoo was
blind because it had had its eyes ripped out by the fishermen who had
caught it. Buckland discovered that it was customary for them to blind
any animal they thought might damage their nets and was horrified:

> If they were spoken to quietly and properly on the subject, I am
> sure that these poor but honest fellows would leave off this horrid,
> cruel custom.[24]

Despite his champion role in the development and regulation of the
fishing industry as an economic sector, Buckland clearly also had a
lively apprehension of the possibility that fish and related animals could
suffer and, no doubt, also had some form of inner life.

Contrasted with this is the astonishing callousness in pursuit of
profit shown to the unfortunate white whales who were exhibited in
the Great New York Aquarium. This was a joint venture between the
showman William Cameron Coup and the wild animal dealer Henry
Reiche. This aquarium was technically well built and had a notionally
educational mission, but it was just a big show in the style of Coup's
mentor Barnum. Its greatest attraction was a massive tank that held two
white whales. These were touted as rare animals otherwise impossible

---

23 Harper, J., *Glimpses of Ocean Life* (London: T. Nelson & Sons, 1860), p. 25.
24 Quoted in Girling, R., *The Man Who Ate the Zoo* (London: Chatto & Windus, 2016), p. 169.

to see. They did not fare well in such cramped conditions and died on a regular basis. But the public didn't know this as on the St Lawrence River, Coup maintained a pen full of white whales (which weren't rare in those days) and as soon as one in the New York tank looked sickly or died it was replaced overnight by a healthier specimen brought down from the St. Lawrence in a specially adapted railway car and the paying public was none the wiser.[25] When Barnum's American Museum and its aquarium burned down in 1865, the whales were burned to death after their tank was smashed in an effort to put out the fire; they then plunged to the street below. They remained, scorched and rotting, on Broadway for several days as they were too big to move.[26]

This book will not explore in detail the controversy around evolution and the essentially theological arguments that raged over the purpose of science and the meanings of scientific discovery. This aspect of the period is one of those things that separate the Victorians from us. As I have pointed out elsewhere, we must always be on our guard against the temptation to see the Victorians as ourselves in frock coats and crinolines. A scientist today may well have a religious faith and may well even believe that in pursuing scientific inquiry he or she is contributing to our knowledge of a divinely created world. However, we are unlikely to find many scientists who see the aim of their work as showing the literal truth of the Bible. In the context of fish and aquariums, one of the most important figures, Philip Gosse, was just such a scientist. His work can only be understood in the context of what we would now call his Christian fundamentalism and, although that doesn't make his natural observations or fieldwork any less valid, it does mean that he understood the significance of these observations and this work in ways which are hard for many of us to understand today. As Rymer Jones said:

> Natural History is the appointed handmaiden of Religion, enabling us to feel and in some proportion to appreciate how

---

25 Mather, F., "White Whales in Confinement", *Popular Science Monthly* 55 (1899), pp. 362–71.
26 Saxon, A.H., "P.T. Barnum and the American Museum", *Wilson Quarterly*, 13 (1989), pp. 130–9.

closely and how carefully the well-being of all creatures has been provided for, – how admirably they are severally adapted to their respective stations and employments, and how wonderfully every part of their economy is made subservient to the general good. This is the true spirit in which the aquarist ought to work, and this is the end and object of his science.[27]

The dates of rocks and the interpretation of stratigraphy often seem in the mid-nineteenth century to manifest themselves as theology as much as scientific disciplines.[28] When Thomas Huxley, the most committed and aggressive of the Darwinian vanguard and known as "Darwin's bulldog", allegedly (but probably apocryphally) said to Samuel Wilberforce, the Bishop of Oxford (who was pouring scorn on the notion of common primate ancestry in a public debate at Oxford University), that he'd rather be an ape than a bishop, we hear the voice of modernity and witness a decisive moment of secular change.[29] And it was in this moment and the controversies that swirled around it and within it that the modern aquarium and the expansion of interest in the aquatic and maritime sciences that accompany it came into being. However, there are many books which explore this issue in detail and

---

27 Quoted as epigraph to Harper, J., *The Seaside and Aquarium* (Edinburgh: William P. Nimmo, 1858).
28 See, for example, Rudwick, M.J.S., *Worlds before Adam: The Reconstruction of Geohistory in the Age of Reform* (Chicago: University of Chicago Press, 2008); Rudwick, M.J.S., *The Great Devonian Controversy* (Chicago: University of Chicago Press, 1985); Cadbury, D., *The Dinosaur Hunters* (London: Fourth Estate, 2000); and Maddox, B., *Reading the Rocks* (London: Bloomsbury, 2017).
29 See Desmond, A., *Huxley, The Devil's Disciple* (London: Perseus Books, 1997) for a full account of Huxley's life and advocacy of Darwinian science. An account of Huxley's debate with the Bishop of Oxford and what he actually said (which is less trenchant but just as amusing) is on pp. 276–81. For his close associate Joseph Hooker, see Endersby, J., *Imperial Nature: Joseph Hooker and the Practice of Victorian Science* (Chicago: University of Chicago Press, 2010) and Desmond, R., *Sir Joseph Dalton Hooker: Traveller and Plant Collector* (Woodbridge: ACC Art Books, 1999). For a sympathetic account of Richard Owen, see Rupke, N., *Richard Owen* (Chicago: Chicago University Press, 2009).

1 The strange case of the missing fish

with much more scholarly authority on the topic than I can muster and I will not be dealing with it except where it is particularly important to the interpretation of a specific incident or object.[30] In a way, every book on Victorian science or related to Victorian science ought to be a book about the relationships of that science to religious controversy and debate. The reader is asked to understand that I am treating the topic of religion as if it were permeating background radiation. As Barbara Gates has pointed out:

> For other Victorians, though not for Huxley, natural history was also akin to religion, since natural history in part stemmed from natural theology – the belief that the discoveries of natural science were proof of the wisdom and power of a divine creation – and was at first indebted to William Paley's 1802 book *Natural Theology*. Throughout the century, natural history continued to be a favorite pastime of clergymen and Evangelicals and a popular subject for the publications of the Society for Promoting Christian Knowledge. On the other hand, natural history became part of the Victorian rage for materialism and material possessions. One could collect birds or eggs or fish and show them off in the home.[31]

---

30 See, for example, Conlin, J., *Evolution and the Victorians* (London: A.C. Black, 2014); King, A.M., *The Divine in the Commonplace* (Cambridge: Cambridge University Press, 2019); and Smith, J., "Eden under Water: The Visual Natural Theology of Philip Gosse's Aquarium Books", paper presented at *Nineteenth-Century Religion and the Fragmentation of Culture in Europe and America* (Lancaster University, UK, 1997) www.personl.umd.umich.edu/~jonsmith/PHGINCS.html. A fascinating micro-history that shows how the reverent approach to natural history could operate in practice is Kennedy, M., "Discriminating the 'Minuter Beauties of Nature', Botany as Natural Theology in a Victorian Medial School", in L. Karpenko and S. Claggett (eds), *Strange Science: Investigating the Limits of Knowledge in the Victorian Age* (Ann Arbor: University of Michigan Press, 2017), pp. 40–61.
31 Gates, B., "Introduction: Why Victorian Natural History?", *Victorian Literature and Culture*, 35 (2007), pp. 539–49, 541.

This book addresses the materialism but it acknowledges the religious context.

Similarly, this book will not explore the ways in which fish and marine science are presented in art and literature. Silvia Granata's recent book on aquariums is especially strong on this as, from 1989, is Lynn Merrill's *The Romance of Victorian Natural History*.[32] Other works with a literary focus are mentioned where they offer appropriate insight. Visual art is, for the purposes of this book, confined to the illustrations to marine science texts or cartoons. The beauties of Ernst Haeckel's *Kunstformen der Natur* (*Art Forms of Nature*) and its Art Nouveau legacy in Anton Seder and Eugène Grasset or the pioneering underwater paintings of Zarh Pritchard or Chris Olsen will not be displayed here.

I thought long and hard about the best way to structure this book and finally decided on a simple chronological account based on the three spatial and institutional domains with some detours. The reason for this is that the progress of aquariums from the Regent's Park Fish House through the craze for domestic aquariums through to the great boom in public aquarium building is, I think, best understood if we observe the different forces at work as they operate sequentially and not thematically. Thus, the reader will find chapters on the Fish House at Regent's Park and the Aquatic Vivarium in Dublin, the domestic aquarium craze, the path-breaking public aquariums at the Crystal Palace and Brighton and the growth of the private and municipal aquariums in simple chronological order. There is a chapter on other aspects of the Victorian encounter with the sea. This looks at marine creatures and marine science in managed nature: seashells and seaweed, angling, fish farming and coastal research institutes. This chapter is designed to enable the various domains of encounter to be understood both in themselves and in relation to each other. Finally, there is an imperial case study where the aquarium movement and some aspects of marine science in Australia will be considered. The focus will be almost exclusively on Britain and the empire and the important aquariums

---

32 Merrill, L., *The Romance of Victorian Natural History* (Oxford: Oxford University Press, 1989).

of the USA, France and Germany will only be mentioned where they throw light on developments in Britain and the colonies.

Although this book presents itself as a history of fish and marine science during the reign of Queen Victoria, it is also a book about animal welfare and one of its aims is to help bring fish into the circle of concern and the argument about animal rights by pointing to the extraordinary cruelty involved in aquariums and the huge mortality among captive fish. As I was revising this chapter there was news footage of Australian rock lobsters (just barred from import into China) being stuffed into plastic crates without a second thought for their comfort. This led me to think of the tanks of pitifully crawling shellfish in the wet market at Wuhan and the slaughter of 17 million mink that had been diagnosed with a mutated form of COVID-19 in Denmark. Lively and energetic creatures kept in alarmingly small cages, the minks were kept for their fur which was mainly destined for China so they would have been slaughtered anyway – it is the enormity of their captivity that horrifies when one already knows that animals will be slaughtered in massive numbers to feed the trade in luxury fabrics. If anyone wanted a practical argument as to why attention to animal welfare and, especially, fish welfare, is in the human interest they could not ask for better examples in the age of a zoonotic global pandemic.

What will also be found in this book is a sustained imagining of a Victorian gentleman or lady or child peering with fascination at a fish swimming in one of the tanks in the London Zoo Fish House in 1853 or wading out into the shallows with a net to catch specimens from a rockpool at Hastings, or drying seaweed and carefully pressing it into an album, or arranging a shell collection and trying to understand the complexities of its taxonomy, or tending to the inhabitants of his or her home aquarium. As we flesh out the picture it will be apparent that what at first seemed a simple image of the past is in fact an extraordinarily complex event made possible only by the confluence of the wider currents of the Victorian world in all its innovative wonder. As I consider the flow of artefacts around the Victorian world – a shell or frond of seaweed picked up from the beach at Hastings, a baby hippopotamus snatched from the southern Nile, a specimen of the giant water lily *Victoria amazonica* from South America or a salmon egg carefully carried across the hemispheres to Hobart, a microscope encountered on a London street or a kaleidoscope on a brass and mahogany stand opening up new visual worlds – I increasingly see something like the internet of things.[33] When I say that Victorians saw fish for the first time I do not, of course, mean this literally. What I mean is that the trivial act of noticing fish was transformed into the profoundly meaningful act of seeing them. This seeing implicated fish and other marine creatures and artefacts into a richly determined complex of both meaning and experience. Fish and related creatures became a thing. The sea became a thing. I'm not here referring to the Heideggerian notion of the "thingness of things" nor even to Bill Brown's "Thing Theory", neither of which I pretend to understand.[34] And while Frank Trentmann's *Empire of Things* uncovers the origins of a consumerist society and, in particular, the development of consumption-based theories of value in the later nineteenth century, I'm less interested in those big social patterns as in "the intelligible universe" derived from Douglas and Isherwood and explored by Asa Briggs.[35] I'm thinking of the transformation of fish into one of those distinctive commodities that enable us to recognise the Victorian when we see it. In the case of other animal products, this has been explored by Anne Colley in her work on animal skins and Sarah Amato in her exploration of the wide variety of ways in which animals and animal products became consumer goods.[36] Emily Cockayne's history of recycling shows us another way of thinking about things and includes examples of the Victorian reuse of fish as glue and other products.[37] These commodities formed a network of the objects that enabled global

---

33 See Simons, J., *Obaysch: A Hippopotamus in Victorian London* (Sydney: Sydney University Press, 2017); Teltscher, K., *Palace of Palms* (London: Picador, 2020); Holway, T., *The Flower of Empire* (Oxford: Oxford University Press, 2013).
34 Brown. B., "Thing Theory", *Critical Inquiry*, 28 (2001), pp. 1–22.
35 Trentmann, F., *Empire of Things* (London: Penguin, 2017). Douglas, M. and Isherwood, B., *The World of Goods* (London: Routledge, 1979); Briggs, A., *Victorian Things* (London: Batsford, 1988)
36 Colley, A.C., *Wild Animal Skins in Victorian Britain* (Farnham: Ashgate, 2014); Amato, S., *Beastly Possessions* (Toronto: University of Toronto Press, 2015).
37 Cockayne, E., *Rummage* (London: Profile Books, 2020).

communication at a time when the Victorians were still learning to be Victorian. And, as Jonathan Crary has argued, they were exploring new physiological knowledge as novel optical apparatus became available.[38] I suggest that the aquarium is one such apparatus and later we will see how the microscope was of crucial importance to the development of marine science as a domestic and leisure activity. Writing in *Household Words* (26 April 1851), James Hannay saw Kew Gardens as "a sort of bank to which botanical currency flows for transmission". Hannay spoke of information as currency, which of course it is, but the key point he makes is that once the currency arrives it is transmitted and not merely deposited.

For the Victorians, the pace of change did not let up and one of those changes was that the fish which had been missing from the lives of their ancestors, except as food, now appeared as the products of a novel complex of forces all tending to the creation of what appears to be a very singular yet commonplace thing: a fish swimming in a glass tank with a human watching it. As an article in *Godey's Lady's Book* (54, 1857) put it:

> It was not only novel but wonderful to behold the creatures of the deep face to face without the aid of a diving-bell, diving-dress, glass-eyes, or the aid of submersion. Who ever thought of taming sea flowers, and jelly-fishes, and crabs and periwinkles?

This book is an attempt to show how that wonder was created and how it was constructed as a commodity within specific institutions. It will be seen that many of the institutions in which fish were displayed failed as businesses and that many people who began marine science and fish keeping as a hobby soon tired of it. So, it could be reasonably objected that the power that I attribute to seeing fish for the first time and the notion of thingness I explore is nothing like as important as I claim. This is clearly a reasonable objection. But I'd argue that transitoriness is also part of the nature of the thing and that novelty, in itself, cannot be a permanent condition. Perhaps the Victorian encounter with fish and the sea is best understood by reference to a sign I once saw hanging

---

38   Crary, J., *Techniques of the Observer* (Boston: MIT Press, 1990).

in one of the fashion halls in the old Kendals Department Store in Manchester:

>You see it.
>You buy it.
>You forget it.

2
The Regent's Park Fish House and the Dublin Aquatic Vivarium

In 1851 two important things happened. The first American edition of Melville's *Moby Dick* was published and the Crystal Palace opened its doors to the Great Exhibition. Although *Moby Dick* would not be recognised as *the* great American novel for many years, the publication of a massive work of fiction with a marine animal at its centre marks a waypoint in the Victorian fascination with fish and the sea. The Crystal Palace may only have had a few fish in ornamental fountains but many people understood that the chief artefact of the Great Exhibition was the Palace itself. It opened many eyes to the possibilities of glass and iron and buildings full of light. So, it was no coincidence that when, less than a year later, the Council of the Zoological Society of London decided to set out on the unprecedented project of creating a collection of fish and putting it on public display, they hit upon a design based on the materials used so triumphantly in the Crystal Palace. A piece in the *Illustrated Magazine of Art* (3, 1854) paid it the compliment of describing it as "a building like the Sydenham wonder of the world of iron and glass". Perhaps more than anything it was the impact of the Crystal Palace that led to the aquarium not only in Regent's Park but also in a thousand parlours and subsequently almost all the major cities of Europe and the USA.

The Regent's Park Zoological Gardens of the Zoological Society of London (henceforth "the zoo" or "London Zoo") was founded in 1828 but, by the late 1840s it was under considerable financial stress and in danger of closure. There were several reasons for this. The British economy was not in good shape and there was not a great deal of discretionary cash around, still less discretionary leisure time. However, although the poor economic context did not help, the bigger problem was that the zoo was run as a private club. Only Fellows (or Members as they were originally called) who paid £2 per year (plus a joining fee of £5), and were subject to a process of recommendation before they could join, could access the grounds. They could bring guests at a cost of 1 shilling each but the combined income from fellowship subscriptions and guest entry did not cover the increasing costs of maintaining a menagerie which was based on the scientific study of animals and birds.[1] Not only was the budget stretched to look after the animals already in the collection but there was little available to buy the new animals that would maintain the zoo's menagerie as a living collection and provide the variety and novelty necessary to extend the zoo's scientific mission and, just as importantly, to keep it in the public eye as a place of interest. The zoo was built round a very different business model from its rival the Surrey Gardens, which maintained a diverse and well-kept menagerie but supplemented this with all kinds of other entertainments such as concerts, firework displays and spectacular performances. And crucially, anyone who could afford the price of a ticket could go to the gardens and would go more than once as new animals constantly arrived and the program of entertainments varied from week to week. It is perhaps indicative of the difference in status of the institutions that when Ensign Cyril Gambier found himself escaping for his life from mutinous sepoys in 1857, he was protected by Indian villagers who inspected him and his fellow refugees:

> as if we were the last [i.e., latest] imported rarity into the Surry [*sic*] Zoologicals.[2]

---

[1] The best recent overview of London Zoo in our period is Ito, T., *London Zoo and the Victorians 1829–1859* (Woodbridge: Royal Historical Society, 2014); for other zoos and menageries, see Simons, J., *The Tiger That Swallowed the Boy: Exotic Animals in Victorian England* (Faringdon: Libri, 2012).

[2] Quoted in Hibbert, C., *The Great Mutiny* (Harmondsworth: Allen Lane, 1978), p. 117.

2 The Regent's Park Fish House and the Dublin Aquatic Vivarium

A junior officer like Gambier would have been familiar with the popular pleasures of the Surrey Gardens but not the elite and academic austerity of Regent's Park.

By 1849 the different fortunes of the zoo and the Surrey Gardens emphasised a number of tensions in Victorian society. Was an institution like a zoo designed for education or entertainment? Could the working class be trusted to comport itself respectably in a place of public resort? How could an institution such as a zoo be harnessed in the service of mass education and, especially, as an organiser of productive and sober working-class leisure? The Fellows of the Zoological Society were doubtful about the answers to all these questions and had been happy to maintain the zoo as an exclusive club where gentlemen with an interest in natural history could observe interesting animals close up. However, within the fellowship there were voices, some idealistic and some pragmatic, that were arguing for change. Some saw the zoo as having a broad mission and proposed that opening it up to a paying public would give it a new lease of life as a place of popular education. Others were unwilling to see the zoo go broke and accepted (whatever their private convictions) that a business model based on a limited number of subscribers was not sustainable and that the best way of increasing revenue was the controlled admission of paying visitors.

This alliance of pragmatists and social reformers won the day and, in 1847, the Zoological Society appointed David Mitchell as its secretary (what we would now call its chief executive). Mitchell understood all the issues that faced the zoo and set about solving the Zoological Society's problems. He pushed through the reforms that would enable more public access to the zoo and, more importantly, he set about the strategic development of the zoo's collection of animals in such a way as to raise more revenue. Equally importantly, he invested the lion's share of that extra revenue in further enhancements not only to the zoo's collection but also to its infrastructure.

The main driver of this strategy was the idea of "star" animals which would so interest people that they would flock to the zoo in their thousands and even pay a supplement to see them. The first of these arrived in 1850. He was Obaysch the hippopotamus and he not only set the zoo's finances at a new level, as Mitchell had hoped, but also established the strategy for the future.[3] Obaysch remained one of the sights of London throughout his life and he was followed by a baby elephant and an American anteater. Both these creatures attracted more visitors but neither of them ever had the same charismatic impact as Obaysch. Nevertheless, Mitchell's strategy was working and the zoo was now able to invest and to create new permanent attractions beyond single "star" animals.

So, in 1850 a decision was taken by the Zoological Society's Council to build something that had not been seen before anywhere – a building dedicated to the display of live fish. *The Morning Post* (3 May 1852) reported on the plan with some excitement and described:

> A house principally constructed of glass and iron for the exhibition of fish, mollusca and other animals, the dimensions of which would be sixty feet by twenty-five feet, which it was expected would enable the Council [of the Zoological Society] to supply in an interesting and effectual manner one of the greatest deficiencies which have hitherto existed in the vivaria of Europe.

This decision was also underpinned by scientific evidence of new discoveries concerning the chemistry of sea water and the solution to the problem of maintaining its freshness. As the aquarist William Lloyd, writing in the *Scientific American* (16 August 1879) almost 30 years later, put it:

> This resolve [to build a fish house] was taken in 1850 and it was the direct consequence of the reading of a paper on the 4th of March in that year, by the late Mr R. Warrington, the chemist to Apothecaries' Hall, London, before the Chemical Society, on the manner in which he succeeded in maintaining some gold fish in unchanged water for a prolonged period.

---

3   For a biography of Obaysch, see Simons, J., *Obaysch: A Hippopotamus in Victorian London* (Sydney: Sydney University Press, 2017). See also the catalogue entries by Helen Cowie in Avery, C., Cowie, H.L., Shaw, S. and Wenley, R., *Miss Clara and the Celebrity Beast in Art 1500–1860* (Birmingham: Barber Institute of Fine Arts, 2021), pp. 92–7.

In fact, Mitchell was so impressed by Warrington's paper that he asked if he might visit Apothecaries' Hall to view his aquarium. This visit took place on 22 January 1852 by which time Warrington had republished his original article, detailing not only some improvements in the design of his freshwater tank but also hinting at the, so far, successful experiments he was conducting with sea water.[4]

Mitchell had also been engaged with James Bowerbank, a member of the Zoological Society's Council, who had managed to keep a colony of sticklebacks alive in a tank sufficiently well for a group (some of whom also kept small tanks) to have formed to study them. As a result of this, Mitchell investigated tank construction and viewed an example made of glass, slate and iron produced by Sanders & Woolcott. This was a company he had employed before to provide plumbing for the recently built Hippopotamus House, and it was subsequently given the contract to develop the tanks in the Fish House.

From Mitchell's point of view, his reading, his research into tank design and his meetings with Warrington and Bowerbank added up to a conviction that what had hitherto been thought all but impossible was now a challenge that, with the right resolve and the right investment, could be met and overcome.

A year later, with the Fish House within a few weeks of opening, the *Morning Post* (4 May 1853) reported on progress with even more excitement:

---

[4] Warrington wrote or delivered a number of seminal papers: "Notice of Observations on the Adjustment of the Relations Between the Animal and Vegetable Kingdoms, by Which the Vital Functions of Both are Permanently Maintained", *Quarterly Journal of the Chemical Society*, 3 (1851), pp. 52-4; "The Aquatic Plant Case or Parlour Aquarium", *The Garden Companion and Florists' Guide* (1852), pp. 5-7; "On Preserving the Balance Between the Animal and Vegetable Organisms in Sea Water", *The Annals and Magazine of Natural History*, 2nd Series, 12 (1853), pp. 319-24; "On Artificial Sea Water", *The Annals and Magazine of Natural History*, 2nd Series, 14 (1854), pp. 412-21. See also Klee, A.J., *Who Invented the Aquarium?*, Aquarium Hobby Historical Society Files, 17 November 2012 (http://www.wetwebmedia.com/FWSubWebIndex/invention_aquarium-1.pdf).

The experiments which had been instituted during the winter and spring, with reference to the conservation of marine and fresh water animals, had been so uniformly successful, that the council had now nearly completed a system of plate glass tanks, in which the greater part of the British zoophytes, crustacean, and molluscs, and considerable number of the fish, would, in the course of a very short time, be exhibited under the most favourable circumstances for observation.

Part of the reason for the uniqueness of this undertaking was that such a display had only recently become possible. At the economic level, glass tax had been abolished and so it was possible to buy large quantities at an affordable price. At the technological level, new developments in glass manufacture meant that large sheets of durable plate glass which were not subject to irregularities and distortions were now available and this made large fish tanks possible. At the scientific level, the chemistry of sea water was now understood as was the mechanism by which it could be kept fresh and aerated so that marine creatures could survive comfortably in something that resembled their natural element. Finally, a combination of science, scholarship and religious fervour had created the beginnings of what would be an overwhelming tide of interest in natural history. And that, together with the discovery of the seaside as a place for holidays (and the idea of holidays at all as the hard years of the 1830s and 1840s gave way to more prosperous times) and the association of the seaside with the new sciences of geology and palaeontology, created a wave of interest and enthusiasm that bore the idea of the Fish House towards a physical embodiment.[5]

---

5   See Armstrong, I., *Victorian Glassworlds* (Oxford: Oxford University Press, 2008); Jackson, L., *Palaces of Pleasure* (New Haven, CT: Yale University Press, 2019); Valen, D., "On the Horticultural Origins of Victorian Glasshouse Culture", *Journal of the Society of Architectural Historians*, 75 (2016), pp. 403–23; Barber, L., *The Heyday of Natural History* (London: Jonathan Cape, 1980); Allen, D.E., *The Naturalist in Britain* (London: Allen Lane, 1976); and Bailey, P., *Leisure and Class in Victorian England: Rational Recreation and the Contest for Control 1830–1885* (London: Routledge, 2006).

2 The Regent's Park Fish House and the Dublin Aquatic Vivarium

The person who brought many of these strands together was the popular author, scientist and Evangelical Christian Philip Henry Gosse.[6] And it was not surprising that it was to Gosse that the Zoological Society turned for advice in setting up and stocking their new Fish House – indeed, Gosse provided over three thousand specimens for the original collection – which opened on 21 May 1853. There was a private viewing the day before where Mitchell gave a talk in which he introduced the fish and set out his ambitious plans for the longer term and his intention to include arctic, temperate and tropical fish in the collection (*Morning Chronicle*, 21 May 1853). We can see from this that the society's plan was to develop the marine collection using the same method as for the collections of animals and birds and to establish it as a comprehensive resource for the study of sea animals. Mitchell also made sure that the Fish House got wider publicity and sent a rough list of the animals on show to Ireland where it was published in the *Dublin Exhibition* (23 May 1853). Reflecting on the Fish House in 1859, the council considered its strategy and the advice and encouragement it had received from, among others, Gosse:

> The attention of the Council having been earnestly directed to the probable success that Marine Zoology might be illustrated in a building specially appropriated to that purpose, an Aquarium was constructed in 1852.[7]

Here we clearly see the council's twin motivations: the advancement of the zoological sciences and the management of a successful exhibition. These by no means conflicting objects underlie almost all its thinking as it developed the zoo in the 1850s and 1860s.

The Fish House was built during 1852 along the lines of a conservatory with large tanks (with a combined length of 84 feet) lining

---

6 See Thwaite, A., *Glimpses of the Wonderful* (London: Faber & Faber, 2002) for a full biography of Gosse, who is perhaps better known as the subject of his son Edmund Gosse's 1907 memoir *Father and Son* where he is portrayed in an unflattering and, I think, unfair, light.
7 *Reports of the Council and Auditors of the Zoological Society of London* (London: Taylor & Francis, 1859), pp. 9–10.

the walls, seven on one side and six on the other, and with smaller tanks down the middle and at each end, standing on what appeared to be cupboards but that were in fact frames containing an ingenious – but alas never used – mechanism that could slightly raise and lower the tanks to create a kind of tidal action. Wilfred Blunt somewhat sniffily and with the considerable benefit of hindsight described the Fish House as "a kind of glorified greenhouse" and the photographs of the building that was constructed bear this out.[8] As we shall see, the design was far from perfect but, at the time, the building made use of the latest cast iron and glass construction techniques – a kinder description might call it a miniature Crystal Palace – and represented what the Council of the Zoological Society and their scientific adviser Philip Gosse believed was the best design for the purpose, bearing in mind that they had nothing with which to compare it.

The building was completed before it was fully stocked (indeed, it was opened before it was fully stocked) and was available to any visitor to see from the outside. The novelty of the building was apparent to William Lloyd when he first saw it. His narrative is worth quoting at length because it gives us a reliable picture of the Fish House as it evolved, shows us the kind of struggle that many men and women of his class (he was working 11 hours per day in a second-hand bookshop and spending what little extra he had in visits to the zoo, where he saw Obaysch in 1850) underwent in what Samuel Smiles aptly termed *The Pursuit of Knowledge under Difficulties*, and also displays the astonishment that the Victorian viewer felt when confronted for the first time with a fish swimming in a tank.[9] On the day of the Duke of Wellington's funeral (18 November 1852), Lloyd was given the day off but instead of watching the Duke's hearse trundle past he walked from his home in St John's

---

8 Blunt, W., *The Ark in the Park* (London: Book Club Associates, 1976), p. 87.
9 William Alford Lloyd is, in many ways, the hero of this book. Yet he is little known. The best source and the source from which the extensive quotations below come in the section on Lloyd are in the excellent website *Parlour Aquariums* which contains an enormous amount of material and was assembled originally by the late Bob Alexander, http://parlouraquariums.org.uk. If any Victorian deserves a proper biography, it is Lloyd. See also Ingle, R., "Who Was William Alford Lloyd?", *Biologist*, 60 (1 December 2013), pp. 24–7.

2 The Regent's Park Fish House and the Dublin Aquatic Vivarium

Square, Clerkenwell, to Regent's Park (about 2.6 miles) as he could only afford the entrance fee to the zoo and not the bus fare as well. Here is his account:

On arriving there I found near the side entrance a building I had never seen before, and which had risen since my last visit – a conservatory-looking glass erection of not large dimensions, standing on a low wall. The door was fastened, and I could see no one inside, and on my asking a passing attendant what the place was for, he said it was "a Fish house, though some people called it an aquarium," and that it was destined to contain fish and other such things, even sea-fishes and lobsters, and that it was intended to be opened in the following spring. He added his disbelief in the success of it, and his sense of the impropriety of its introduction in a zoological garden.

He regarded it as evidently an innovation on the customary inhabitants of such a place, which he defined as "beasts, birds, and reptiles." I was impressed by the novelty of the idea, however, and went away (after again trying to get in to see Mr Gould's collection of stuffed humming birds, shown in the other (northern) side of the Garden in what is now the Parrot House ...)

So after looking longingly at the Sixpenny Garden Guide Books on sale in the same room, and wishing I could afford to buy one, I went back to the Fish House, and passed round to the rear, and there to my great astonishment I saw through the glass side of the building a glass tank containing perfectly clear water, with some aquatic plants growing in it (I have since learnt to call the plant Valisneria) and, wonder of wonders, a living pike!

I wish I could write what I then felt; I wish I could now feel as I then felt, but such freshness of wonder comes to one not more than perhaps half a dozen times in a life. I could not get away from the place (it was at the extreme north-east corner of the building, and the tank has been years ago converted into a marine one) and remained there until it began to grow dusk and it was time to get home.

In returning I thought of all the odds and ends I had read of fish against my will, of the "Vesica Piscis," of Moule, of Lucy, and of Crooked Lane, and on reaching St John's Square I made a compound meal of dinner, tea and supper in one, consisting of eggs and bacon and potatoes and tea. But it was not eggs and bacon and potatoes and tea – it was Pike and Valisneria and water! I HAD SEEN AN AQUARIUM, AND THAT WAS ENOUGH. I never felt unkindly towards the old archaeological pursuits, but here was something fresh, and green, and living, without any Dr. Dryasdust about it, and I had seen amidst living vegetation a living pike, with his gorgeous livery of mottled green and gold.

We should not only enjoy this voice from the mid-nineteenth century as it speaks directly to us but also because we learn three very interesting things from Lloyd's account.

The first is that before the Fish House opened it was already being referred to as an aquarium and although it is more or less universally assumed that it was Gosse who suggested the name and coined the word, in fact Mitchell had, before Lloyd's visit, met Warrington and seen his paper "The Aquatic Plant Case, or Parlour Aquarium" in the January 1852 issue of *The Garden Companion and Florists' Guide*, so even though this name was advocated for by Gosse it may well have been in Mitchell's mind before Gosse suggested it. Gosse did, however, wisely point out that aquarium was to be preferred to aquatic vivarium, which he wrote was a name of "awkward length and uncouthness … unsuitable for a popular exhibition or domestic amenity".[10] Even so, as late as 1868, The Reverend R.W. Fraser was still able to offer instructions for the establishment of a "Marine Vivary" in the old style with anemones, molluscs and crustacea rather than fish as the main inhabitants.[11] And an intriguing possibility has recently been raised

---

10 Gosse, P.H., *The Aquarium: An Unveiling of the Wonders of the Deep Sea* (London: John van Voorst, 1854), p. 250.
11 Fraser, R.W., *The Seaside Naturalist* (London: Virtue & Co., 1868); this was the third edition of Fraser's earlier book *Seaside Divinity* (London: J. Hogg, 1861) which in turn built on his work for young people *Ebb and Flow: The Curiosities and Marvels of the Sea-shore* (London: Houlston & Wright, 1860). Interestingly, the chapter on aquariums was not in *Seaside Divinity* which suggests that Fraser or his publisher were aware of the commercial enhancement offered by adding the new material. On Fraser, see Moore, P.,

2 The Regent's Park Fish House and the Dublin Aquatic Vivarium

by Emily Senior. She points out that, when Gosse was in Jamaica in the mid-1840s, he knew about George Johnston's experiments in maintaining a balanced seaweed and invertebrate aquarium and recorded in his journal that he had set up something similar for marine worms. She points out that in his book of drawings, *Jamaican Studies 1844–1846*, he sketched a "fountain aquarium" which is almost identical to the one depicted in *The Aquarium* some 12 years later. When he returned from Jamaica he already had, fully formed and visualised, the idea that would soon become the classic aquarium and had already had experience of creating one.[12]

The second is that the problem of keeping the tanks fresh and clear had not arisen where this pike is concerned. There are probably two reasons for this. Firstly, the water plants in the tanks may have been sufficient to maintain a healthy balance of the sort that Warrington had described and demonstrated – these plants were carefully prepared by having their root wrapped in a small ball of clay and sand that was then itself wrapped in canvas and the whole then planted in the pebbles and gravel on the floor of the tank. Secondly, the freshwater tanks were always kept better than the marine tanks and the arrangements by which water flowed through them were – even at this early stage and before they were connected to the steam-driven system that circulated water around the gardens – sufficient to keep at least one fish healthy.

Finally, we learn that the pike was happily swimming in its tank some five months before the Fish House opened and was kept healthy all that time. This is borne out by the photograph taken of it by the Comte de Montízon in 1852, which was the first ever photograph taken of a live fish swimming in water.[13]

"Seaside Natural History and Divinity: A Science-inclined Scottish Cleric's Avoidance of Evolution (1860–1868)", *Archives of Natural History*, 40 (2013), pp. 84–93.

12 Senior, E., "'Glimpses of the Wonderful': The Jamaican Origins of the Aquarium", *Global Currents*, 19 (2022), pp. 128–52.

13 The Comte de Montízon was the Bourbon pretender to the thrones of both France and Spain. He was offered the throne of Mexico but sensibly declined. He was a founder member of the (Royal) Photographic Society and, among other things, produced an important series of images of animals in Regent's Park. Photography was obviously attractive to displaced royalty. Maharajah

The all-important water was treated in different ways, depending on whether it was fresh or sea water. Although the solutions adopted by the Zoological Society were only partially successful and seem very crude when compared with the immense engineering projects that made large public aquariums possible later in the century, they nevertheless speak to the ingenuity and scientific innovation that characterised the enterprise under Mitchell's leadership. The large freshwater tanks running down each side of the building were supplied from the same source as the gardens generally and this water flowed through them and then into sunken tanks whence it drained away. The small freshwater tanks were not connected to this flow and were simply changed with new fresh water when it was required. The sea-water tanks were kept fresh by a series of siphons and tanks which culminated in a large cistern high up and cooled by the fresh water flowing into the freshwater tanks. From this cistern the water dribbled back into the tanks, which were thus aerated and freshened twice per day. The smaller marine tanks were, like the small freshwater tanks, simply given new water as occasion demanded. This system was extremely expensive to run as the sea water was costly to acquire and the relatively inefficient purification system coupled with the lack of temperature and light control meant that huge amounts had to be constantly replaced and much which could have been saved and recycled was wasted.

The Fish House was part of the first wave of the new infrastructure program funded by Mitchell's business plan and, by 1851, projects realised or in train included a hippopotamus house and tank (for Obaysch), an antelope house, an aviary for eagles, the massive

---

Duleep Singh, the heir to the Sikh Kingdom and rightful owner of the Koh-i-Noor diamond, was also a member of the Society. See also Martinez, A., "A Souvenir of Undersea Landscapes: Underwater Photography and the Limits of Photographic Visibility", História Sciéncias Saûde-Manguinhos, 21 (2014), pp. 1024–41. The first genuine underwater photograph was taken by William Thompson in 1856. He constructed a special metal box for his camera and activated it by pulling a string. Predictably enough the photograph was taken at Weymouth, a town at the heart of the marine science craze. The first underwater photographs showing any detail were not taken until 1893 when the French photographer Louis Boutan constructed a flashlight system capable of illumination in even quite deep water.

hummingbird collection displayed by the master taxidermist John Gould as part of the zoo's successful strategy to attract some of the huge throng of visitors who had flocked to London to see the Great Exhibition, and the Fish House itself. The 1852 edition of the *Popular Guide* to the zoo promised:

> A building erected in the spring of 1852 for an Aquatic Vivarium, in which will be exhibited both fresh-water and marine fish, mollusca and other animals.[14]

In its earliest manifestation, the Fish House's collection seems quite modest when we consider the enormous variety of exotic marine life that is expected in a modern aquarium. Down the centre were saltwater tanks containing sea anemones. On both sides the tanks were filled with fresh water and contained a selection of common British fish: pike, perch, roach, tench, dace, rock bass and trout. A more exotic specimen was the long-eared sunfish. The Fish House was always intended to offer a complete picture of marine life, so at one end of the new building stood a tank for diving birds and, at the other, a tank for wading birds. In the first season there were 58 species of fish, 75 of molluscs, 41 of crustaceans, 27 of coelenterates, 15 of echinoderms and 14 of annelids in addition to the British freshwater fish mentioned above, so although there was great variety there was still a preponderance of things that crawled or scuttled or just stayed put rather than swam. It is worth noting that while the zoo itself was a display of the exotic that metonymised the reach of empire, the Fish House was, at this point, quite different and reflected the life of local rivers, rockpools and coastal waters. So, as well as being a new departure as far as the class of animals on display was concerned, the Fish House was also a significant modification to the more general mission of the Zoological Society.

The *Athenaeum* magazine (21 May 1853) reported on the opening and described six tanks containing crustacea, molluscs, echinoderms, nudibranchs and sea anemones. At this point in the history of the display of marine life it is often the case that things like crabs and

---

14 Mitchell, D.W., *A Popular Guide to the Gardens of the Zoological Society of London* (London: The Zoological Society. Printed for the author, 1852), p. 22.

anemones are more commonly encountered than fish. I think this is, in part, because of the first successful aquarium experiments which used madrepores and, in part, because of the relative ease of collecting and transporting this kind of creature as opposed to fully grown live fish. And it is partly because, under the tutelary influence of Gosse, marine life was imagined much more in terms of the beach and the rockpool than the open sea and the Fish House was, in its arrangements, in many ways a realisation of the littoral marine science which, at this point, held the public imagination in thrall. But to see anything in this new setting was still astonishing and, as witnessed by the *Athenaeum*:

> The aquarium immediately became a fashion, a rage, an infatuation, which, now that we are sobered down and are able to regard a stickleback with equanimity and a sea anemone without any sensible increase in pulsation it seems difficult to realize.

The Fish House was not, in fact, an especially important or successful exemplar of the public and private aquariums we will encounter in this book and the reasons for this will be dealt with later in this chapter. However, it established that new Victorian experience: seeing a fish swimming in water. The Gradgrindian-sounding *Year-book of Facts in Science and Art* for 1853 summed it up well:

> In this fish-house we can see at a glance what days of watching elsewhere could not afford. The ferocious pike has become docile as a puppy; the perch – always invisible amongst the deep holes of the rivers and lakes here yields up the secret of his haunts. ... Ere long every inhabitant of London will be able to see what up to the present time has been seen only by the adventurous and sea-towed dredger who, casting his net to the bottom of the ocean has beheld its numerous inhabitants in the freshness of life ... A new world of animal life will be drawn from the depths of the ocean for the amusement and instruction of the present generation.[15]

---

15 Timbs, J., *The Year-book of Facts in Science and Art* (London: David Bogue, 1853), p. 228.

2 The Regent's Park Fish House and the Dublin Aquatic Vivarium

Punch (25 June 1853), as it often did, exactly caught the mood:

> Who can contemplate the Marine Vivarium at the Zoological Gardens without congratulating himself on living in an age when the public derives that amusement from zoophytes, which, at a former period, it would have sought in bear fights?

Here we see the main themes that structure the discourse made possible by this new institution: the Fish House. It is both entertaining and educational and so conforms to the mission of the modern and competitive zoo envisaged by Mitchell. It offers an experience indoors that hitherto was only available to hardy individuals outdoors (much like seeing a hippo swimming in a tank in the middle of London offered an experience only previously available to travellers in dangerous parts of Africa). Finally, it offers a new encounter and a way of seeing a class of animals which became part of the novel experience of urban tourism with books such as Peter Cunningham's *Handbook for London* or Edward Mogg's *Mogg's New Picture of London* showing visitors – who may never have been to London before – what to see and when and how to see it and how much to pay for it. Thus, the spectacle of fish nests within a bigger spectacle: that of London itself seen in fresh ways and through new eyes where the act of looking becomes, in itself, a desired, structured and organised experience with a specific economic value.[16]

Seeing fish was a true novelty and, if there is nothing else of value to be found in this book, I would contend that the connection with the Victorian mind – and understanding how exciting it was to see a fish and how cutting-edge that experience was – makes the whole enterprise worthwhile. The French scholar Philippe Hamon, writing about nineteenth-century exhibitions, called the aquarium "un objet incontournable de la culture visuelle du XIXe siècle" ("an unmissable object of nineteenth-century visual culture") and the centrality of the new institution to new ways of seeing can be readily identified across a range of manifestations.[17] But it was in the Fish House that this

---

16 Cunningham, P., *Handbook for London*, 2 vols (London: John Murray, 1849); Moggs, E., *Moggs's New Picture of London* (London: E. Moggs, 11th edn, 1848).

experience was first made possible and it is thus a site of novelty and innovation the importance of which cannot be underestimated.

The opening attracted a good deal of media coverage and we can sense the excitement in this description of that event in the *Morning Chronicle* (21 May 1853):

> In these [the tanks], to speak in plain language, a variety of seaweeds and plants, of the most various and beautiful descriptions, rise from a soil of marine pebbles, shells and rocky and mossy fragments. Within the recesses of these appear all manner of strange fishes – marine sticklebacks, crabs of every kind, and some with periwinkle shells upon them, scorpion-looking prawns by scores, strange molluscae, with their lungs on the outside, baglike creatures which attach themselves to the said periwinkles, and live upon sucking their shells; strange animals too, like bunches of living macaroni; ugly fishes all out of shape; other ugly little fishes with no shape at all – all swimming creatures, adhering creatures, crawling creatures, creatures with horns, creatures with great eyes, creatures sparkling in brilliant suits of scales, creatures of a clammy white – all these moved or hung in their different fashions to suck such substances as suited them – water-weeds, stones, and sometimes like suckers clung to the smooth plate glass of the tank.

The Times (31 May 1853) gave its readers a similarly excited account (syndicated from the *Literary Gazette* which, being a somewhat more upmarket publication than the *Morning Chronicle*, didn't hesitate to throw some Latin around):

> A living exhibition of the sea-bottom and its odd inhabitants is such an absolute novelty, that we must give our readers this week, at the risk of being charged with undue partiality for natural history, some account of the elegant aquatic vivarium just opened to the public. On the borders of the flower-bed in the Zoological

---

17 Quoted in Le Gall, G., "Dioramas Aquatiques: Théophile Gautier visite l'aquarium du Jardin d'Acclimatation", *Culture et Musées*, 32 (2018), p. 82.

2 The Regent's Park Fish House and the Dublin Aquatic Vivarium

Gardens, Regent's Park, has been constructed, crystal-palace fashion, of glass and iron, a light airy building sixty by 20 feet in area, containing around its transparent walls fourteen six-feet tanks of plate glass. Eight tanks will, in the first instance, be devoted to living marine animals, and of these six are ready for exhibition. They enclose masses of rock, sand, gravel, corallines, sea-weed, and sea-water; and are abundantly stocked with crustacean, starfish, actiniae, shelled and shell-less mollusks, and fish of the genera *gasteroseus, labrus, crenilabrus, blennius, gobius,* and *cattus*. The whole are in a state of natural restlessness, now quiescent, now eating and being eaten.

The piece then goes on to capture the very specific novelty of the Fish House experience:

One of the most interesting features in this novel exhibition is the restless change of position among the several creatures. The visitor may occupy the whole day in passing in and out of the building from tank to tank and yet every time see something new.

We know how fish behave in a tank or, through wildlife documentaries, underwater. But the Victorians didn't. As *Household Words* (9, 1854) put it:

There [in the Fish House] we may look at a new world that has been lately spread before the eyes of men, and begin – as even naturalists have only within this last year or two – to pick up an intimacy with the little people of the sea.

Three years later, the novelty was still worth reporting on:

Around us we perceive slices of the deep-sea bed and the rapid river ... There is a noble pike lying still on a stone – a model sitter for the photographer who lately took his portrait

(*Quarterly Review*, 98–9, 1856)

The Fish House also appeared to address one of the most profound problems facing Victorian zookeepers. How did you get people to make repeat visits when you had much the same animals? A successful commercial zoo like Belle Vue in Manchester made its menagerie just a part of an enormous variety of entertainments; Regent's Park did it by constantly acquiring new animals. But here we also see a mode of exhibition that is never the same twice. Once you've seen a lion, you've seen a lion. But a tank full of fish, always changing, always mobile, is a different matter entirely. And certain of the fish were especially valued as entertainment. The conger eel, for example, would "come out to be fed at a signal from the attendant in the Fish House" while the hermit crab was "one of the most amusing and restless animals in the aquarium". The shrimp, however, was "not an attractive animal in the aquarium as it moves about but little".[18] There was always a tension in the zoo between the scientific mission of the Zoological Society and the need to develop the gardens as a space of polite and edifying entertainment. This was the commercial reality of the enterprise so, while the Fish House was, perhaps, the most obviously scientific of the displays on offer, even here it was acknowledged that the animals had to earn their keep, if only by moving about and eating from time to time.

Hogg's Instructor (June 1853) pointed out that the "exhibition of live fish" was "still wanting its finishing touches" but there was enough to show to thrill those lucky enough to see the tanks in the earliest days and to offer them yet another new thing at the zoo. It will be noted from these accounts of the Fish House as it was when it opened that you could see the fish both by going inside the building and by walking round the outside. The journalists also draw attention to the construction of the building and to the plate glass forming the tanks. So, what the Fish House offered was also the opportunity to experience new technologies and, as we shall see, in accounts of the large public aquariums that opened 20 years later and other aquariums and marine exhibitions going forward to the end of the nineteenth century, the technology was an important part of any aquarium. This sometimes, especially when it was deployed at large scale, created as much interest

---

18 Holdsworth, E.W.H., *Handbook to the Fish House in the Gardens of the Zoological Society of London* (London: Bradbury Evans, 1860), p. 29.

2 The Regent's Park Fish House and the Dublin Aquatic Vivarium

and excitement as the fish themselves. In fact, these aquariums should be considered not only as centres for marine science but also as testing grounds for innovation in engineering.

Notice also that, at this stage, fish were a relatively small part of the display. This was partly, as I suggested before, because of the influence of Philip Gosse and the desire to re-create the tidal rockpools the study of which (indeed, the existence of which) he had introduced to a public desperate for a new craze. The Fish House thus offered the pleasures of visiting the Gossean seaside and capitalised on the frenzy for littoral science. But from this derives also the creation of apparently natural environments for the various creatures to swim or crawl around. These replicate the features of rockpools and the commentators cited above pick up on the importance of the pebbles and gravels to the display. Amy King has pointed out that Gosse's motivation in stocking the aquarium was "a realist representation" of the seaside.[19] And the analogy between the aquarium and the rockpool was still being made by Fraser in 1868.[20]

My guess is that the zookeepers knew very well that the public was seeing the animals through eyes already accustomed, by Gosse, to think of the rockpool as the site of scientific inquiry and to look on science as a leisure pursuit on a seaside holiday. It is also worth noting that the construction of realistic environments was, in some regards, a matter of life or death for the animals and the attempt to create such contexts for their captive lives contrasts strongly with what was going on in the bare cages and featureless yards in the rest of the zoo. It was also, more practically, because the sourcing of fish was a hit-and-miss business and transporting them in good condition was an even riskier activity.[21]

---

19   King, A.M., *The Divine in the Commonplace: Reverent Natural History and the Novel in Britain* (Cambridge: Cambridge University Press, 2019), pp. 138–9. See also King, A.M., "Tide Pools", *Victorian Review*, 36 (2010), pp. 40–5 and King, A.M., "Reorienting the Scientific Frontier: Victorian Tide Pools and Literary Realism", *Victorian Studies*, 47 (2005), pp. 153–63.
20   Fraser, R.W., *The Seaside Naturalist* (London: Virtue & Co., 1868), p. 242.
21   The illustrator and naturalist Sarah Bowdich Lee, for example, had samples of live fish sent through the post sealed in bladders. They did not always survive the journey. See Gates, B., *Kindred Nature* (Chicago: University of Chicago Press, 1998), p. 78.

So creating environments that optimised their chances of survival was good business as well as good science. Whereas the Zoological Society could mount an expedition to capture a particular animal (or use its network to do so as in the case of Obaysch the hippopotamus), it was not possible to guarantee that a particular fish would pop up in a dredge or net. Indeed, the early British ichthyologists John Reay and Francis Willughby derived many of their more unusual specimens from the fish markets of Venice and elsewhere and thus their pioneering efforts in a new science proceeded on a necessarily serendipitous path.[22] As the *Handbook to the Fish House in the Gardens of the Zoological Society* put it in 1860:

> Many curious animals are frequently transported to the gardens but have too often suffered so much on their journey, as to be unable to survive under the unnatural conditions, which, to a certain extent, must exist in the best managed Aquaria.[23]

And, as we shall see, the transportation of fish in healthy condition remained a problem for many years.[24]

---

22   Birkhead, T., *The Wonderful Mr Willughby* (London: Bloomsbury, 2018). Reay is perhaps more commonly known as Ray and his *De Historia Piscium* (1686) (actually a collaboration with Willughby – perhaps more commonly known as Willoughby – who had died in 1672) has the distinction of causing a delay to the publication Sir Isaac Newton's *Principia Mathematica* because the cost of producing it and its failure to attract buyers used up the Royal Society's budget. Fortunately, the astronomer Edmund Halley stepped in and bankrolled Newton's book out of his own pocket and it was published in 1687. No sooner had this happened than the Royal Society told Halley that it could not afford his £50 salary and that it would pay him in unsold copies of *De Historia Piscium*, which must have made him reflect that no good deed goes unpunished.

23   Holdsworth, E.W.H., *Handbook to the Fish House in the Gardens of the Zoological Society of London* (London: Bradbury Evans, 1860), p. 2.

24   In fact, the first successful transportation of live fish from the Southern Hemisphere did not happen until September 1868. The fish came from the Seychelles and got as far as Paris where they died. See Wright, E.P., "On the Transportation of Living Fish from South of the Equator to Europe", *The Annals and Magazine of Natural History*, Ser. 4, 2 (1868), pp. 438–41.

2 The Regent's Park Fish House and the Dublin Aquatic Vivarium

To some extent this accounts for the apparently conservative, even dull, collection that was first offered to the public. The *Illustrated London News* (28 May 1853) spoke of:

> The ingenious purpose of exhibiting, under natural conditions an amazingly curious and beautiful division of the animal kingdom which, hitherto but little known to the popular observer, people the hidden regions of the lake, the river and the sea.

The sensational nature of the Fish House may be gauged by its appearance in George Cruikshank's cartoon "Passing Events or the Tail of the Comet of 1853".[25] This is a collection of things of note that happened during the year and these were catalogued by the novelist Cuthbert Bede (real name Edward Bradley) in his description of a visit to Cruikshank's studio.[26] Among the jumble of little figures is the American anteater (Mitchell's star animal for the year) and behind it a tank clearly marked "Vivarium" and containing an eel, a crab and two fish.

The problem of maintaining aquariums, especially when there is sea water in the tanks, is not trivial. It was as recently as 1847 that Anna Thynne had managed to sustain a sea-water aquarium capable of supporting the collection of madrepores which were the object of her scientific interest. Part of this breakthrough was her discovery of the chemistry of sea water and, especially, the role of seaweed in maintaining a healthy and clean marine environment. By 1849, her collection – certainly the first modern aquarium – was available for viewing by scientists in her house (the sub-dean's residence in Westminster Abbey) and was suitably marvelled at. Anna Thynne was one of the first of the long procession of Victorian women engaged at the leading edge of marine science that we will meet in this book and who are, by and large, poorly remembered, if they are remembered at all.[27]

---

25 Cruikshank, G., "Passing Events or the Tail of the Comet of 1853", *Cruikshank's Magazine*, 1 (1854).
26 Jerrold, B., *The Life of George Cruikshank*, 2 vols (London: Chatto & Windus, 1894), I, pp. 317–18.
27 Stott, R., *Theatres of Glass* (London: Short Books, 2003) offers a full account of Thynne's life and contributions to marine science.

The Fish House or Aquatic Vivarium, or sometimes Aquavivarium, also soon changed its name and became the Aquarium. The idea of a curated and sustainable collection of marine animals for public exhibition is familiar to us and almost every town of any size has a Sea Life Centre or its equivalent. But there had never been anything like this before and the word "Aquarium" seems to have been waiting for the exact moment when David Mitchell cut the tape and let the public in on that day in May 1853. Just as Philip Gosse had advised on the construction and stocking of the Fish House, so he now gave the Zoological Society a new word that perfectly fitted the thing it had brought into being. It is usually said that Gosse coined the word and the *Oxford English Dictionary* gives only nineteenth-century citations post 1854 but in fact it was around as early as 1807 when an advertisement in *The Times*, placed by a Mr Tod who was selling plans for hothouses, included "an aquarium" (most likely meaning a bowl or tank used for growing and displaying aquatic plants) among the possible projects available. Tod claims that he had recently built things for "various noblemen and gentlemen" but, alas, doesn't tell us whether any of these commissions included an aquarium. It was probably in this sense that Robert Warrington had been using the word at least as early as Gosse. *Curtis's Botanical Magazine* (78, 1852) had an article on the "aquarium" at Chatsworth and this certainly referred to a container for aquatic plants only. The name was viewed with some suspicion in the *Eclectic Review* (9, 1855):

> The word "aquarium" will never, we hope, be added to the number of big and bad, dark and dead words which disgrace the literature of natural science.

The writer preferred the plain English-sounding "tank" (which is, in fact, derived from a loan word into Portuguese from Gujarati or Marathi). But perhaps a piece entitled "Sea Views" in Charles Dickens' periodical *Household Words* (9, 1854) got it right:

> The lodgings provided in the Regent's Park for the small people of the sea – first called the Aquavivarium and now the Marine Aquarium – for a new thing there was a new name wanted, and

2 The Regent's Park Fish House and the Dublin Aquatic Vivarium

the first name is not always the best – have given satisfaction to their tenants.

It may well be that this view of the satisfaction of the fish was overstated but the author is surely right in identifying the serendipitous choice of the word "aquarium" and also in understanding that here a new thing had come into being. And this had implications for every other thing that made up the Victorian world as the grammar of their relations with each other and with people now had to change to accommodate the novelty.

People had kept fish in various ways for a long time but what was new in the mid-nineteenth century was the ability to sustain a healthy marine environment in the home and to keep a greater variety of fish alive for a longer time. At least, that was the idea. As we will see, many would-be aquarium keepers and, indeed, public aquariums, found the task of maintaining fresh sea-water daunting.

The Fish House was initially a huge success. There were 40,000 visitors in the first week. As a comparison, Obaysch the hippo, by far the greatest attraction ever seen at the zoo, attracted an average of 16,281 extra visitors per week in his first 14 weeks on show. The Annual General Meeting of the Zoological Society held on 2 September 1853 was told that in August there were 20,748 more visitors than in August of 1852 and most, if not all, of these were coming to see the fish. And, in celebration one feels, a Mr Salmon was proposed for the fellowship. Receipts went up to £17,500 compared with £12,800 in 1850. The secretary was able to write in the *Literary Gazette* (47, 1854) of the collection of:

> fish, molluscs, zoophytes, and other aquatic animals, first projected in 1851, which has probably excited more attention from the novelty and the intrinsic beauty of the objects themselves than any other of the recent additions to the collection.

And this was after the cult of Obaysch had established itself. The former secretary of the society wrote to Edward Charlesworth in March 1854 giving him good news about the state of the zoo and telling him that

> The Fish House is bringing more money than did the Hippopotamus. Last year's receipts exceeded any former year – except the Crystal Palace year – and 1854 at present promises to exceed 1853.[28]

As the *Daily News* (21 May 1853) accurately observed when the Fish House opened:

> Those who formerly ran after the hippopotamus may now perhaps find equal amusement in the more remote but not less curious living novelties of this new glass house.

And so they did. As *Fraser's Magazine* (1854) put it in its article "The Aquarium":

> Learned and unlearned, young and old, may there be seen gazing at patent wonders hitherto concealed in the bosom of the deep.

Here the object of vision is not only the inhabitants of the tanks but also the people looking at them. What is being reported on is not just a collection of marine animals in an innovative display but the new experience of seeing a fish made possible and constructed by the coming together of all the elements that made the Fish House capable of being realised both as a physical entity and as an existential domain.

And let us pause to consider just what 40,000 visitors in a week means. Let us assume that everyone who came to the zoo that week wanted to see the Fish House and let us assume that the visitor number was divided equally over seven days (in fact, it would not have been, as Sunday remained the exclusive province of Fellows and their guests); that means that roughly 5,714 people visited the zoo each day. In 1852 the zoo was open from 9 am to 'sunset'. In late May in London sunset is round about 8.30 pm. Let's assume that the keepers started moving people out no later than 8 pm and that those who arrived at 9 am were able, somehow magically, to transport themselves straight to the Fish House. That gives

---

28 Ito, T., *London Zoo and the Victorians 1829–1859* (Woodbridge: The Royal Historical Society, 2014), p. 120.

a practical window of opportunity for visiting of 11 hours which means that an average of about 520 people would have been in the Fish House during any given hour. In a building not much more than 60 feet long and of proportional width and with much of the floor space taken up by the tanks, that makes for a pretty intimate experience. And although this is a calculation which is smoothed by a number of unrealistic assumptions, the impact of those assumptions (an equal number of people each day and an equal number each hour) is to disguise what must have been at times an almost unbearable crowd and an experience mitigated only by the weather: May 1853 was moderately cold with a mean temperature of only 50° F so being inside, even in a crush, must have been a relief – especially to poorer visitors who may not have had warm coats. Visualising what a visit to the Fish House in that first week might have been like helps us to do two things. Firstly, we can better understand the various newspaper reports of the opening. Secondly, we are reminded that if you wanted to see something in Victorian England you had to go and see it in person and that, however mediated an experience might become, the initial encounter was always existential.

And it is that experience, rather than necessarily the animals themselves, which is described in many early accounts of the Fish House, such as this from *The Leisure Hour Monthly Library*, (3, 1854):

> What a rich vista is this opposite which we are now standing! How crowded with living beings – fishes, crustacean, mollusca and zoophytes! Here are submarine rocks and grottoes, arches, caverns, and recesses, all festooned with trailing fronds ... What a strange romantic picture it is of the depths of the sea, called up, as it were, by the potent wand of a magician, to bewitch our senses and half persuade us that we really are in the calm profound, surveying all around us, as a traveler resting in some rocky dell.

The author here is describing the novelty of the Fish House experience for those who have not had it and may never have it by deploying the terms of familiar Romantic literary discourse and the pictorial conventions that accompany it. He is forced to express the uniqueness of this new thing so falls back on familiar tropes but the hyperbole is telling and eloquent of the problem facing any author who wishes to describe something that he has just seen himself for the first time to a reader who has not seen it nor anything like it except in pictures that were, to an extent, guesswork. As Jonathan Smith has observed:

> Until the development of the aquarium made the aquarium-view part of Victorian culture's iconography scientists were reluctant to depict underwater scenes for fear of being labeled fanciful or speculative.[29]

This aquarium-view developed very quickly: as early as 1 January 1854, *The Illustrated Magazine of Art* published 16 engravings of specimens from the Fish House and as these were disseminated so the experience of seeing a fish was mediated for a much wider public than the one that was able actually to visit the House itself, large though this was.[30]

And, of course, the aquarium-view was not only a contribution to iconography but also a contribution to the way Victorians experienced the world and their place in it.

Queen Victoria loved going to the zoo and she was also interested in fish and fishing. She was an early visitor and on 15 July 1853 she recorded:

> A showery morning, but I went all round the Zoological Gardens with Bertie. We saw some very interesting animals. Nothing is more interesting than the Aquarium, and the molluscs etc one sees there, feeding, are most extraordinary.[31]

---

29 Smith, J., *Charles Darwin and Victorian Visual Culture* (Cambridge: Cambridge University Press, 2006), p. 3.
30 Interestingly, Joseph Wolf's de luxe production *Zoological Studies Made for the Zoological Society of London from Animals in Their Vivarium, Regent's Park* (London: Henry Graves & Co., 1861–1867) did not include any fish. This is probably partly because fish went beyond Wolf's artistic comfort zone but it also reflects the semantic problem regarding a distinction between animals and fish.
31 Queen Victoria's journals are now fully digitised and available to search. They may be found at www.queenvictoriasjournals.org.

2 The Regent's Park Fish House and the Dublin Aquatic Vivarium

Notice that she uses the word "Aquarium" and not "Fish House" or "Aquatic Vivarium" or "Aquavivarium". Here is the queen of the most modern and urbanised country on earth with a modernising prince consort enjoying seeing a new word on the page. The queen bought a print of the photograph of the pike for her album of zoo photographs.[32]

Punch (22 October 1853) testifies that the popularity of the Fish House had not faded after a few months of being open:

> I had nearly forgotten to mention that one Sunday afternoon I paid 6d for a 'bus to Regent's Park Zoological Gardens, and 6d. to come back again without seeing anything. Professor Jelly of the Institution says the Vivarium is very interesting; but I find it is only the aristocracy who are admitted on a Sunday; the working-classes it seems would do an injury to their souls by looking at the fish on that day.

This is a contribution to a debate that was raging at the time and not just about the zoo. The question was: should public buildings such as museums be open on a Sunday and what entertainments, if any, should be available when this involved a charge for admission? A strong Sabbatarian movement mounted a powerful and largely successful campaign to keep things closed. But, as *Punch* points out, private institutions such as the Zoological Society were permitted to open for members only. Thus, in spite of the liberalisation of access to the paying public which, thanks to Mitchell's management, had given the zoo its financial lifeline, there were still limitations – unfair according to *Punch* – which prevented people from seeing the most popular exhibition in the zoo on a Sunday. And, as we shall see, the public and private aquarium movement fought a rolling brawl with the Sabbatarians well into the 1870s.

The success of the Fish House is, perhaps, most effectively demonstrated by the fact that not long after it opened *The Times* was carrying an advertisement for the opening of an aquarium at the Surrey

---

32 This photograph was shown as Montizon's contribution to the Society of Arts' first photographic touring exhibition in 1854.

Zoological Gardens. This Marine Vivarium was also noted in the *Morning Post* (24 August 1853) and had:

> fine large glass tanks procured especially for this collection by P.H. Gosse, Esq. ... and forming the most instructive and amusing object ever yet presented to the public.

So, whatever the strengths and weaknesses of the scientific endeavour there was no doubting the popularity of the enterprise and this clearly justified the not inconsiderable investment in tanks for a rival aquarium. Notice that Gosse was also advising the proprietors of the Surrey Gardens – when the Zoological Society discovered this, they disputed and reduced his fee, as they were already not happy that he had mounted his original expedition to collect specimens before a contract had been agreed and then simply presented them with his bill. But Surrey's choice of Gosse is hardly surprising as, in 1853, there were very few people indeed who understood the principles of aquarium management. As we shall see in subsequent chapters, the development of aquariums across Europe, America and the British empire in the next two decades was made possible by the advice of a relatively small number of experts who moved from institution to institution as opportunities became available.

Rather than attempt to create artificial sea water, those responsible kept the tanks supplied with water that was shipped in. This caused problems of its own and the "London Correspondent" (probably David Mitchell) of the *Inverness Courier* (26 May 1853) reported that the casks in which the water was transported were having to pass customs where they were checked for brandy. The water collected was shipped in on an Antwerp steamer. *Fraser's Magazine* (50, 1854) explained:

> The marine tanks are supplied with sea water ... taken up halfway between the Scheldt and the Thames where the water is supposed to be purest. Sea-water from Dover and Brighton becomes speedily turbid, perhaps in consequence of its being taken at an insufficient distance from the shore.

2 The Regent's Park Fish House and the Dublin Aquatic Vivarium

And with raw sewage discharged directly into the sea being the norm at this time, one would certainly prefer to get water from an offshore source. Later, the London Zoo aquarium was supplied with water from the Bay of Biscay. However, although every care was taken, this arrangement was not as successful as was hoped and huge amounts of sea water, which was very expensive because of the need to transport it, was wasted as the Zoological Society could not keep it pure. William Lloyd bought hundreds of gallons of the waste water and by a process of filtration and aeration brought it back to health so it could be re-used in his own aquarium business.

Many things combined to send the Fish House into a period of crisis and decline. The main problem was in the design. Keeling puts it very simply:

> Throughout its existence, which lasted into the early decades of the twentieth century when its large tanks were converted into the Diving Birds House, its staff had to grapple with temperature problems. For half the year the exhibits were too hot, and too cold the other half. This problem was caused because the house was constructed from that finest of all conductors – glass – one of the advantages, or disadvantages, of early pioneering and experimentation.[33]

To understand just what a problem the Zoological Society had on its hands we can see readings taken by William Lloyd:

> On 13th July last (1859) at four o'clock in the afternoon. The thermometer in the House stood at 93° F, and in the Tanks, with a free bulb immersed in the water, it was 82° F. Nothing in the shape of animal life ... can long resist the destructive effects of heat such as that.[34]

---

33 Keeling, C.H., "Zoological Gardens of Great Britain", in V.N. Kisling, ed., *Zoo and Aquarium History* (Boca Raton, FL: CRC Press, 2001), pp. 49–74, p. 70.
34 Lloyd, W.A., *A List with Descriptions, Illustrations and Prices, of Whatever Relates to Aquaria* (London: W. Alford Lloyd, 1858), p. 159.

If we want to understand the impact of rising sea temperatures on ocean life, we could do worse than study the history of the Fish House.

In an era when only shade, ventilation and hand-fanning could offer relief, it was hardly surprising that the Fish House got hot. And especially when one considers how many people may have been crammed in at any one time. The Crystal Palace offered an example of the management of high temperatures through a combination of calico sheets and an elaborate system of ventilation louvres, manned by soldiers of the Royal Engineers, which were opened and shut on the basis of a two-hour cycle of monitoring of 14 thermometers within the building. This kept the heat at only a few degrees above the outside temperature. But even so there were days when going round the exhibition was a sticky experience. Interestingly, two other innovative glass buildings, the Winter Garden in Regent's Park (1845) and the Palm House at Kew (1848), had also had heat-control problems which had been solved but, at the zoo, no one seems to have learned from these examples.

It was an irony that the cutting-edge technology, the design and the materials employed, which so impressed the public, were the worst choice for a collection of this kind. However, although the design of the building was fatally flawed there were also problems in the curatorial regime and the resources needed to sustain such a large aquarium. As Blunt put it:

> But in fact the fish had never had it so bad, for the fish-cum-plant alliance [i.e., the combination of animal and plant life needed to maintain a balanced aquarium] was not as simple a matter as it had first appeared. If there was insufficient light, the plant died and polluted the water; if the water was changed too frequently, the fish resented it and sickened; and if there was too much light then the fish died. So at the Zoo there were men constantly pulling blinds up and down, and hand-syringing the water to regulate the supply of oxygen ... The time eventually came when the Zoo also found the difficulties of keeping a satisfactorily balanced aquarium too great to justify the labour and expense involved ... By the early seventies the fickle public had, in any case already lost

2 The Regent's Park Fish House and the Dublin Aquatic Vivarium

interest, and the Vivarium – henceforth curtly dismissed as "the old Fish House" – was closed.[35]

Even so, in 1854 the council reported that in addition to a "beautiful group of young salmon" being added to the collection by the kind donation from Lord Ranelagh:

> The collection of fish, zoophytes and mollusca, has fully maintained the interest which it excited at first.[36]

This report added that the whole enterprise must be of "satisfactory method and design as several animals of the original collection are still alive". But one wonders if this declaration was the beginning of a recognition that all was, in fact, far from satisfactory. Certainly, although the first edition (1854) of Gosse's *Aquarium* praises the Fish House because several of the animals had lived there for nearly a year and some sea water had not required replacing for seven months, these positive words were omitted from the second edition in 1856. And, indeed, as early as 1856 the council was being very frank about the problems (while still acknowledging the enormous success of the Fish House in creating new interest at the zoo):

> The Aquarium has continued to afford the greatest gratification to visitors; and extended experience in its management will soon, it is hoped, obviate some of the difficulties which have hitherto interfered with the perfect exhibition of the animals which inhabit sea-water.[37]

Lloyd, who would probably have had inside information, knew that the council was aware of the issue much earlier:

---

35 Blunt, W., *The Ark in the Park* (London: Book Club Associates,1976), p. 88.
36 *Reports of the Council and Auditors of the Zoological Society of London* (London: Taylor & Francis, 1855), p. 18.
37 *Reports of the Council and Auditors of the Zoological Society of London* (London: Taylor and Francis, 1856), p. 12.

The conservatory-like building now standing in the Gardens, was discovered to be in the first summer of its existence, an arrangement so utterly wrong, that the modifications it demanded would amount to something like an entire reconstruction.[38]

In successive reports one can trace the efforts of the council to solve the problems of managing such a complex exhibit. The glass was painted an opaque white to try to cool the building (and thus one of its major attractions – the fact you could see many of the fish from the outside as well as the inside – was compromised). By 1861, the necessity to buy "new canvas for the fish house" at the not small cost of £5 16s 2d (presumably for shade) was noted in the report for that year. In 1863 a new tank costing £6 5s 5d was installed. Even so, on 25 October 1865, *The Times* wrote critically of the poor facilities at the zoo (especially the cramped cages) and singled out the Fish House as an example:

> The Fish-House too, is nothing like large enough, even for the common inhabitants of the sea and fresh-waters.

This appears harsh when one compares it with the rapturous eulogies that were the commonest way to describe the Fish House in 1853 and one assumes that the experience was now too commonplace as were the "common" fish who lived there.

The problems caused by the design of the building, and the great difficulty of maintaining its temperature coupled with the relatively crude methods of keeping the water healthy, led to significant mortality among the fish and the growth of thick green and purple vegetation which choked up the tanks and turned the sea water in particular a dark brown to the point where it was all but impossible to see through it beyond three inches – which made it difficult to view such fish as managed to stay alive. Mitchell tried to make the best of this and in his report for 1854 he observed that:

> Algae are growing luxuriantly in those tanks which are not agitated by the vivacious evolutions of the sea-fish, and this

---

38 See Parlouraquariums.org.uk, *Parlour Aquariums*.

secondary feature is well worth the attention of the botanist, to whom the opportunities thus afforded of studying the development of these plants are of the most complete character, while the extremely beautiful effect of the colour, dependent partly on the Algae themselves, and partly on the peculiar action of transmitted light, are not less instructive to the artist.[39]

A year later he described the lobster:

who comes straggling over the stones in such an ungainly manner, [and] is more like a moving salad than any living thing, so thickly are back, tail, feelers and claws infested with a dense vegetable growth.

(*Quarterly Review*, 98, 1855–56)

Even so, it is noteworthy that in 1856 another essayist, also in the *Quarterly Review*, noted that "the water remains pure and bright". Admittedly, this appears to be a comment on the freshwater tanks which were always in better condition than the marine tanks but clearly there were different perceptions at work.

The sad state of affairs that had developed and the failure of repeated solutions were finally addressed by the Council of the Zoological Society in its report for 29 April 1868 when it recorded:

A resolution to build a new fish house. The present Fish House (built in 1852) was the first exhibition of fishes and other animals living in water ever erected. It was built at a date when the necessary requirements for preserving animal life in aquaria were very imperfectly understood. Since that period many other similar buildings have been erected at various Gardens on the Continent, of larger dimensions and on much improved principles such as the Fish Houses in the Jardin d'Acclimatation and in the Zoological Garden of Hamburg and the Aquarium House at Hanover.[40]

---

39 *The Zoologist*, 12, (1854), p. 4278.
40 *Reports of the Council and Auditors of the Zoological Society of London* (London: Taylor & Francis, 1868), p. 18.

And although the society had clearly understood where it had erred and had models of the kind of thing it now wanted around Europe (and subsequently in the USA), the experience and expense of the Fish House failure meant that the new aquarium, originally planned for 1913 and postponed for obvious reasons, was not built until 1923 and was opened in 1924. Sea water was to be collected in the Bay of Biscay and shipped to the zoo on barges down the Regent's Canal. Its design and attractiveness were such that it didn't close until 2020 when the Zoological Society finally gave up on fish in Regent's Park and moved its aquatic collection to a conservation-oriented aquarium at Whipsnade, its rural site.

In order to maximise the chances of success of any future aquarium, the Zoological Society even supported overseas enterprises. In 1874 it made a grant of £100 to the new Zoological Station in Naples (the first marine science research institution). This was not a purely altruistic gesture. As the council report for that year put it, the grant was made:

> Firstly, in view of the benefits likely to come to Zoological science from its establishment, and, secondly, in expectation that valuable acquisitions to the Society's Fish House (which the council hope shortly to see rebuilt on a much more extended scale) would ultimately be received by means of this Institution if it were once set on a permanent footing.[41]

However, in spite of the challenges, the Zoological Society persisted for a few decades and the annually published *Guide* to the zoo recorded, throughout the 1860s, 1870s and 1880s, new acquisitions: axolotls in 1868, "a Darter, the only one ever obtained alive" in 1875 and, as late as 1886, there was still an impressive variety of fish listed in the saltwater tanks.

In the first few years of its existence, the uniqueness of the Fish House meant that the members of the public were not as aware of its shortcomings as they would be later when they had seen other

---

41 *Reports of the Council and Auditors of the Zoological Society of London* (London: Taylor & Francis, 1875), p. 8.

2 The Regent's Park Fish House and the Dublin Aquatic Vivarium

aquariums. Indeed, when a new star animal (the anteater) came on the scene in October 1853, the Fish House still held the public in its thrall and even Obaysch took a back seat:

> The Fish House alone, of all the attractions in the Gardens, maintained its position against the new-comer. The unique and beautiful collection of living forms there displayed will long constitute one of the chief sources of amusement and instruction the Gardens contain, and is little likely to lose its interest whatever other additions the place may receive.
> (*Hogg's Instructor*, ser. 3, vol. 2, 1854)

There is a reason for this beyond the uniqueness of the collection and the experience it offered. If you were a Victorian, the only possible way you had to see the fish was to visit the Fish House. You could read about them, you could hear stories about them but you could not see them without actually going to Regent's Park and looking at them. We are so used to all manner of visual representations and virtual simulacra that we need to remind ourselves of the existential component of the Victorian encounter with fish – and many other things – where mediation played a secondary role to "being there and doing that".

By 1856 a poem in *Punch* (14 June 1856) shows that the Fish House was an established part of the London scene when it is viewed through the eyes of a young lady apparently lamenting the dreadfully wet weather that summer:

> Then dear Mr Mitchell's Vivarium,
> The pleasantest refuge I know
> While we're kept in this constant Aquarium
> (As Frank says) how is one to go?

Interestingly, June 1856 was part of the longest dry spell on record in England (August 1853–July 1856) so one must assume that there had been one or two wet days and the young lady is exaggerating. One of the many good jokes in *Punch* but all but impossible to spot more than 160 years later.

It is noteworthy that, while the Fish House was struggling almost from the outset, the venture still did well in other ways. The *Handbook to the Fish House in the Gardens of the Zoological Society* published in 1860 was still a substantial publication listing many species and pointing out things of special interest, such as the fact that the conger eel "will often come out to be fed at a signal from the attendant in the Fish House" or that it was "necessary to cover the tank [in which the pike lived] as on more than one occasion a specimen has jumped out during the night". This *Handbook* remained in print for many years but never went into a second edition and thus was unlike the *Guide to the Gardens* which was updated annually as new animals were acquired and new buildings constructed. Its value was more as a history of the enterprise and a textbook of marine science. This perhaps shows how unstable the collection was and how transitory the lives of some of the fish. The necessarily haphazard modes of collection are perhaps best witnessed by the award of the society's bronze medal in 1869 to William Penney in recognition of his many donations to the Fish House that came in without any obvious system or plan.

The guide *London and Its Sights* (1859) described the vivarium as one of the "chief objects of interest" in the zoo (together with the hippopotamus and the rattlesnakes), but, only two years later a similar venture, *The Visitors' Guide to Places Worth Seeing in London*, did not mention the fish as one of the special attractions.[42] The 1862 edition of Cruchley's *London Guide* recommends a visit to the Fish House on the grounds that it has:

> given rise to many interesting discoveries in their [fish] habits and economy.[43]

In 1870 the *Guide to the Gardens* was still able to claim that:

---

42 *London and Its Sights* (London: T. Nelson & Sons, 1859); Philp, J.M., *The Visitors' Guide to Places Worth Seeing in London* (London: Ward & Lock, 1861).
43 Cruchley, G.F., *London Guide, A Handbook for Strangers, Showing Where To Go How To Get There and What To Look At* (London: G.F. Cruchley, 1862), p. 267.

2 The Regent's Park Fish House and the Dublin Aquatic Vivarium

The success which attended this experiment, then first publicly attempted on a large scale, has assisted in promoting the popular study of these most interesting creatures in a very remarkable manner not only in this country but also on the continent.[44]

This summary, although certainly true, may be contextualised by the balanced, if critical, reflection on the Fish House that appeared in the *Illustrated London News* roughly a year later (2 December 1871):

Although we regard the Regent's Park aquarium as having set a good example in the means of observing many animals which, previously to 1853, were known in life to but a few naturalists; and although, by the simplicity of its construction and by its general arrangements (as then understood, however, erroneously), it stands in marked contrast with the pretentiousness of the character of many of the French and German aquaria just named, yet it was soon discovered to be very defective. In particular, it was made at a period when it was the fashion to imitate the very successful iron and glass building of the Exhibition of 1851, and, accordingly the Regent's Park aquarium was constructed like a conservatory or hothouse, set on a low wall of masonry. But, however well such an erection might be adapted for some forms of vegetation, and creatures needing much light and warmth, it was the very worst possible for a collection of British aquatic plants and animals, the primary conditions of the existence of which are shade and coolness. Consequently, in the first summer of its existence, the mortality of the animals in the Regent's Park aquarium was very great, and the vegetation was stimulated into far too rapid a growth, which rendered the water very turbid. The modifications which have since been made in the building, and which still exist there, have but partially remedied its original defects, among which have also to be enumerated the very serious ones of the small dimensions and tall and narrow proportions of the tanks (thus giving insufficient air-absorbing surfaces of

---

44 Sclater, P., *Guide to the Gardens of the Zoological Society of London* (London: Bradbury, Evans & Co., 1870), p. 42.

water); and chiefest of all, the very serious fault (in any but very diminutive tanks) of the absence of adequate means of purification by keeping the water ever in motion, as in the sea, and in rivers, and even in ponds; this motion being needed in addition to the aeration effected by plants growing in the water.

So just over a quarter of a century into its life, the Zoological Society could describe the Fish House as an experiment and its shortcomings were sufficiently well known to be forensically scrutinised in a respected magazine. However, we often learn most when experiments fail. More to the point, although it may have proved a daunting task to keep going, its success was not in the outcomes for the House itself but rather in its role as a stimulant to further scientific study. And to show this was not simply the post hoc rationalisation of a potential embarrassment on the part of the Zoological Society, we may remember that when the House was first opened, basic scientific equipment was made available to viewers. For example:

> A microscope was screwed to the bench adjacent to the *Serpulae* so that they [visitors] can observe ... the motion of these *cilia* and their transparency [which] makes them beautiful objects.
>
> (*Fraser's Magazine*, 50, 1854)

The sudden appearance of a new scientific tool created value to studies all over the empire. An article by E.F. Kelaart on Ceylonese pearl oysters published in the *Madras Journal of Literature and Science* in 1857 – and it's surprising that in that year people had time to publish articles on oysters given everything else that was going on in India – pointed out "the great facility an *aquarium* gives for the investigation of the natural habits of molluscs" and that research could now reach new levels because of this invention. Kelaart had ordered some tanks from England but, in the meantime, he had had some custom-made in Ceylon from glass, slate and concrete; you wonder what the local artisans made of the commission.[45]

---

45 Kelaart, E.F., "Introductory Report on the Natural History of the Pearl Oyster of Ceylon", *Madras Journal of Literature and Sciences*, 3 (1857), pp. 89–105.

2 The Regent's Park Fish House and the Dublin Aquatic Vivarium

In 1877 an attempt was made to restore the Fish House and the society appointed William Saville-Kent, a noted marine scientist who had been involved in the design and management of the large public aquariums at Brighton and Manchester. By this time the tanks had filled with chalk (presumably due to limestone deposits from the London fresh water) and there were very few fish to be seen. He estimated that this would be a three- to four-month job and asked for a salary of £300 pro rata. He was offered only £50 for the project but to supplement this he was also offered the permanent half-time post of director for £200 per annum. He turned this down but did take up the project and used his position to network support for a proposed marine research station in Jersey as he had concluded that public aquariums, even ones as prestigious as the Fish House, could not be effective sites of scientific research and, having done what he could to get the tanks into better shape, he moved on.[46]

If the Fish House was an experiment, it was a noble experiment and, in 1882, it was again repaired. At this time a large tank for auks, guillemots and penguins "under water" was added. Certainly, by this time it appears that the marine display was far less important than the display of diving birds which also captured the same novel experience that the original tanks had nearly 30 years earlier. So where in 1853 the astonishment was caused by "stirring scenes" such as a big trout eating a smaller one, now one could be amazed by seeing a penguin glide by, to all intents and purposes flying through the water.

The Times (5 December 1899) published a kind of obituary for this once much-loved institution that had started with such hope and innovation but that, after nearly half a century, was now at the end of its life:

> The fish-house – Gosse's Aquavivarium – has been practically reconstructed, and other entrances provided. No attempt is now made to keep seafish or lower marine organisms for the exhibition of which the house was erected; and it is now known as the diving birds' house.

---

46   Harrison, A.J., *Savant of the Australian Seas* (Hobart: Tasmanian Historical Research Association, 1997), pp. 32–33.

At the very end of our period, *The Spectator* (23 November 1901) painted a very bleak picture:

> The fish house is chiefly used as feeding-place for the diving birds. It is dark, cold and depressing, and the very few fish kept there do not thrive, neither is at all suited for the confinement of shore birds, which live in a rather darkened damp space at the end opposite the diving birds.

In 1903 the Fish House was formally renamed the Diving Birds House, which appears to have been an official recognition of the informal name it had had for several years.

It would take 20 years and a suite of technical advances before any further experiments were made in large-scale aquarium-keeping in England, although in Europe, especially in France and Germany, the example of the Fish House inspired several new aquariums which were often developed using the expertise of Englishmen, like Mitchell and Lloyd, who had experience of the Fish House. But, as we shall see in the next chapter, the Fish House inspired many hundreds, possibly thousands, of people to set up aquariums in their homes and the experience of seeing fish was thus transferred from the setting of a formal public and highly prestigious scientific institution to the drawing room, parlour and study. And a whole new industry grew up to support the craze.

Although the Fish House stimulated more emulation on the Continent and in the United States than in Great Britain, one institution did set up in its shadow and that was the scarcely acknowledged Aquatic Vivarium in Dublin Zoo. In order to maintain a chronological narrative, I will deal with this interesting institution now.

The Royal Zoological Society of Ireland (RZSI) began to plan for an aquarium in 1853, which makes me wonder if the note that was sent to Dublin by David Mitchell in May of that year was rather more than a simple piece of the publicity surrounding the opening of the Fish House. The Council of the Society noted that although "some measures have been taken", the accomplishment of a finished aquarium was not easy "owing to the large supply of good water, soft and fresh, required". Fortunately for the society, the International Exhibition (the

2 The Regent's Park Fish House and the Dublin Aquatic Vivarium

Irish equivalent of the Great Exhibition) between May and October 1853 brought over a million extra visitors to Dublin and many of these went to the zoo. This led to a 50 per cent increase in the zoo's income which enabled it to invest in various projects, one of which was the aquarium:

> They have built a house for the display of aquatic animals, and have had a forcing pump, tanks and pipes etc fitted for a due supply of water to this interesting addition ... They have also erected a windmill, with a view to aerating the water in the several fish tanks, fifteen of which, it is hoped, will be in action within the next few days.[47]

The troubles of the Fish House appear to have led the RZSI to develop its building on a very different plan to the Fish House; the members of the society must have had many contacts in London and the zoological world more generally that enabled them to learn from the problems that beset Regent's Park. They appear to have done a better job with pumping and aeration. As for temperature issues, in the first winter the problem was frost, which caused minor damage to the building. The principle of the Dublin Vivarium was constant aeration of the water and the great champion of this method in later years, William Saville-Kent, attributed this to Robert Ball, the secretary of the RZSI.[48] This method contrasted with William Alford Lloyd's approach via permanent circulation which required massive storage tanks and was, therefore, more expensive to set up and needed much more space. However, it is worth noting that at this early stage, when there were only two aquariums in the world, they had respectively adopted the two distinctively different principles on which the design of aquariums

---

47 Royal Zoological Society of Ireland, *Proceedings of the Society, 1840–1860* (Dublin: The Council of the Society, 1908), p. 81.
48 Saville-Kent, W., "Aquaria: Their Construction, Management, and Utility", *Journal of the Society of Arts*, 24 (1876), pp. 292–8, p. 293. On Robert Ball, see also Jackson, P.N.W., "Robert Ball (1802–1857): Naturalist", *The Irish Naturalist's Journal*, 30 (2009), pp. 15–18. See also De Courcy, C., *Dublin Zoo* (Wilton: The Collins Press, 2009) which is the official history and contains material on the Aquatic Vivarium including the photograph in this book.

would be based for the remainder of the century (albeit on a very small scale in Regent's Park). According to John Taylor – and this was later confirmed by Lloyd who said he had seen a leaflet advertising this facility – the aeration system at Dublin was also organised by the unique ploy of installing hand pumps and inviting visitors to take a turn as part of their Aquatic Vivarium experience.[49] There is a possibility that this system was activated by the provision of a small tip to the keeper in the Vivarium. The encouragement of visitors to give gratuities to staff was very much part of the culture of Dublin Zoo and this may be borne out by entries in the society's minute books for December 1860:

> Pump damaged and no gratuity to be given till the pump is restored or the perpetrator discovered.

And January 1861:

> Limiting gratuities to the men because of the pump.

Presumably with a hand pump out of action, they fell back on the windmill.

The frost damage was soon repaired and the society took the opportunity of enlivening the display still further:

> Another novelty has been introduced in the large slate tank, mirrors are concealed in the angles next to the spectators, and throw light on the fishes from below and on the rock work with which the tank is lined, the consequence is that the most minute details of the fishes are rendered visible, and at some distance from the tank these animals appear, as it were, to give out light, and thus attract attention when they could not be seen at all under the old construction.[50]

---

49   Taylor, J.E., *The Aquarium: Its Inhabitants, Structure, and Management* (London: Hardwicke and Bogue, 1876), pp. 15–16.
50   RZSI, *Proceedings*, pp. 89–90.

2 The Regent's Park Fish House and the Dublin Aquatic Vivarium

So, while the Fish House was ablaze with light, the Aquatic Vivarium was clearly dark and, although this greatly helped to maintain the temperature at a reasonable level and to inhibit the growth of algae and weeds in the tanks, it made the fish hard to see. It may be the addition of this borrowed light that stimulated a comment in the 1856 *Proceedings of the Natural History Society of Dublin*:

> You are aware of the extreme interest excited by the study of the Vivarium where the habits and progression of animals may now be scanned, whose ways were previously hidden in the deeps of the ocean.

In 1856 the council of the RZSI considered that it had been successful in the management of "Aquatic Vivaria" [*sic*] and was planning "a considerable increase in the department" that included a further development of the aeration system. This paid off and by 1859 the council could report that:

> The marine and fresh water aquaria are in perfect order, and continue to present a series of objects of great interest.[51]

It is surely noteworthy that, from the earliest days, the society had succeeded in managing both types of aquarium (notice also that the council was now referring to what had been the Aquatic Vivarium as an aquarium) and that this success continued to stimulate the popularity of the zoo and the numbers of visitors, which is frankly acknowledged in the report of 1860:

> The marine and fresh water aquaria have been well supplied throughout the year with fishes and invertebrate animals, and continue to be attractive to the public. Many curious marine animals, which otherwise would be scarcely known to anyone but the exploring naturalist, are by the aquaria rendered familiar to thousands of delighted visitors.[52]

---

51 RZSI, *Proceedings*, p. 108.
52 RZSI, *Proceedings*, p. 116.

The report added that all this was achieved at a "very trifling expenditure", which speaks to the efficiency of the pumping and aeration systems and which enabled the expensive sea water to be kept clean and pure and to require no great volumes of replacement. It is worth noting that, by 1860, the mania for home aquariums in England had reached its high point and was now in sharp decline. So, the contents of the Fish House would have been far more familiar to the English public. But it appears that things were different in Ireland and the public facility provided by the zoo was still the place where most people would encounter fish for the first time and be amazed by them.

While the council in London realised very early on that a successful aquarium needed a building on a very different plan to the Fish House, it was unable to set aside the requisite resources for many years. The RZSI, however, started to plan for a new aquarium building in 1867. This was opened in 1870 and reflected not so much the problems encountered in the original building, dark though it may have been, but rather the desire to capitalise on its success:

> The arrangements made by the Council for the filtering of the sea water supplied to the new Aquarium are now completed, and when the return of Spring shall enable the Council to stock the sea water with fish and zoophytes from Dalkey and Howth, it is expected that the Aquarium of the Dublin Gardens will yield to none in Europe, in the variety and beauty of its interesting inhabitants.[53]

In 1870 this was probably a justifiable claim to make but, as we shall see, the examples of the large European aquariums meant that institutions on a scale that would dwarf the Dublin facility were in planning in England and would open their doors very shortly.

Nevertheless, the aquarium continued to thrive and to be improved: new lighting and tanks were installed in 1886 and skylights added at each end in 1887. Clearly, the problem of gloom never quite went away but, nevertheless, the Dublin Zoo aquarium showed that it was possible to maintain healthy tanks and healthy fish over a long

---

53   RZSI, *Proceedings*, p. 14.

2 The Regent's Park Fish House and the Dublin Aquatic Vivarium

period and demonstrated that, when this was achieved, a display of fish and other marine animals would always be a popular attraction and thus a commercial asset to any zoological garden.

Unfortunately, there is very little archive material relating to the Dublin aquarium and even the notes of the RZSI Council meetings are completely lost for some years and only available in reconstructed and reported form for others. So, the greatest mystery of the Aquatic Vivarium will probably never be solved. And that is – who did the RZSI call in for advice when they were planning and establishing their facility? We know that they had received advice on what to put in the tanks from David Mitchell, but who advised them on design and other technical matters? The various reports of the Zoological Society of London (ZSL) and archived letters for the relevant years show that there was plenty of correspondence between London and Dublin on zoological matters but, alas, there is no evidence from those sources of specific collaboration on the development of the Aquatic Vivarium.

The obvious person that the council members in Dublin would have turned to was Philip Gosse but there is no record of him having worked for them. Yet it seems unlikely that the council would have contemplated such a risky capital project without expert support. At the time the Aquatic Vivarium was being built, Gosse was still disputing with the ZSL about payment owed for his efforts in stocking the Fish House so he may have needed some extra cash. On the other hand, the RZSI's report makes it very clear that some preliminary work had been done on the project as early as 1853, so, if Gosse had been the adviser, this would have been at much the same time as he was helping to set up the Fish House. Finally, the design of the building and aeration system in Dublin was very different from that adopted in London. Did Gosse have the capacity to recommend such a radically new arrangement? Although he was an expert in theory and a fine fieldworker, his own success as a practical aquarist was not outstanding and his views on water purification not entirely sound. A consideration of the names of the RZSI Council members and donors in the relevant years sheds no light on the matter. So, we have to conclude on the available evidence either that the RZSI set up the facility based on its own analysis of the science and the tips from Mitchell or that Gosse was called in but all record of this interaction is now lost.

I incline to the view that, despite the risks, the council went back to first principles both in building design and in its evaluation of the science regarding water aeration and out of its deliberations and arguments came the success that attended its Aquatic Vivarium. This is a likely possibility because at the time the Aquatic Vivarium was planned and developed, the secretary of the RZSI was Robert Ball, an amateur naturalist so keen that he had been pensioned off from his job as a civil servant because of the time he spent on his scientific interests. Ball had serious interests in marine science and had connections with some of its most prominent exponents. His sisters were the eminent phycologist Anne Ball and the entomologist and conchologist Mary Ball, whom we will meet again in a subsequent chapter, and he himself was the inventor, in 1838, of a tool called "Ball's Dredge", which is used even today in slightly modified form for collecting marine specimens. I think that, given his position and his interests and those of his immediate family over many years, it is highly probable that much of the impetus for the Aquatic Vivarium came from him and that he may well have been able to deploy his experience as a scientist to help work through the decisions that needed to be made to ensure that the enterprise would flourish.

3
The domestic aquarium

> Is your home furnished with that never-failing source of entertainment – an Aquarium? If so, purchase Dean and Son's 1 shilling manual on the subject.

This interesting question was posed – and answered – by an advertisement in *The Times* (17 March 1856). But notice that the advertisement is aimed at people who already have an aquarium. Now they need a book to show them how to look after it.

The opening of the Fish House created not only the sensation of being able to see fish swimming clearly in replicas of their natural environments, but it also stimulated one of Victorian England's periodic crazes. By 1857 the *Athenaeum* could refer to the aquarium as "a household institution" and a year later the *National Magazine* (3, 1858) pointed out that:

> From the very first day that the tanks were exhibited at Regent's Park, the experiment was everywhere repeated ... We shall not exaggerate greatly if we say that everybody now has an aquarium, we see them in the parlour windows of quiet streets, in the halls, drawing-rooms and conservatories of the wealthy, and they form very attractive features of many of the leading exhibitions.

And the *Literary Gazette* (21 August 1858) wondered that:

> It was almost if we went to bed one night innocent of anything but ... having seen some glass tanks at the Zoological Gardens, and rose to find every naturalist's shop, half the fishing tackle houses, and all the filtered water and ginger pop establishments displaying elegant assortments of living, swimming fishes ... while more than half the centre drawing-room windows in the most fashionable parts of town appeared furnished with an ornamental chest of plate glass.

This last observation really catches, I think, the astonishing novelty of the experience of seeing fish up close, which the Fish House provided for the first time.

The *Telegraph* (13 April 1858) claimed that:

> No tasteful home is now without an aquarium.

And the *Saturday Review* (4, 1857) considered that the aquarium "has taken a place in our homes". Writing with the benefit of experience and hindsight, William Lloyd considered that:

> This institution [the Fish House], and it only, which now seems so small and insignificant gave an impetus toward what became from about 1855 to 1860 a real and impetuous aquarium mania all over Britain. (*Scientific American Supplement*, 16 August 1879)

The *Illustrated London News* (30 December 1871) made the relationship between the Fish House and the home aquarium craze even more explicit:

> The exhibition of the above-named Aquarium in Regent's Park gave enormous impulse to the popular study of marine and fresh-water animals and plants; indeed, it may be said to have originated the movement which commenced in 1854 and which for a few years assumed the character of a "mania" that an aquarium in every house became quite an institution; and small ones, complete with

3 The domestic aquarium

glass, water plants, and animals, were, in 1855, hawked in the streets of London and sold at a very small price.

Everyone wanted to keep fish, to have a little piece of the sea in their parlour. Some wished to do it simply to be entertained by the fish. Others were more serious about the study of marine science and endeavoured to replicate the scholarly mission of the zoo in their own home. Here the anonymous author of "A Fisherman's Sixth letter to his Chum in India" in *Bentley's Magazine* (42, 1857) makes a direct connection between the Fish House and his decision to set up a home aquarium:

> The last time I went to the Zoological Gardens I spent my whole time among the fishes. It was there that the idea struck me that I might beguile many a weary hour making a collection of fresh water fish.

In 1856 the Fish House may have been already seen as a problem within the offices of the Zoological Society but, in the outside world, this doesn't appear to have been widely recognised. So the difficulties of maintaining a fresh and healthy marine environment on a large scale, which more or less defeated the Zoological Society, were to become the challenges of hundreds, maybe thousands, of dogged aquatic enthusiasts some of whom made considerable investments in pursuit of this or other hobbies. As Charles Kingsley somewhat sniffily wrote in *Glaucus*:

> Your daughters have, perhaps, been seized with the prevailing "Pteridomania," and are collecting and buying ferns with Ward's cases wherein to keep them (for which you have to pay), and wrangling over the unpronounceable names of species (which seem to be different in each new Fern-book that they buy) till the Pteridomania seems to you somewhat of a bore: and yet you cannot deny that they find enjoyment in it, and are more active, more cheerful, more self-forgetful over it, than they would have been over novels and gossip, crochet and Berlin-wool.[1]

---

1 Kingsley, C., *Glaucus: or, The Wonders of the Shore* (London: Macmillan & Co., 1855), p. 1.

Notice how Kingsley assumes that fern collectors will be women (and, as we shall see, aquarium keeping and marine biology seem to have been similarly gendered) and that a father (or husband) will be meeting the not-inconsiderable outlay for Wardian cases as, one assumes, he would have to do for the similar aquariums. In the era before the *Married Women's Property Act*, this was, except for widows and single women of independent means, the inevitable model. It is also worth noting that many fashionable houses already had the "chest of plate glass" in the drawing-room window mentioned by the *Literary Gazette* in the form of Wardian cases of ferns and miniature conservatories packed with greenery.[2]

But while the experience of the Fish House may have retarded the development of other large-scale public exhibitions, its effect on the small-scale domestic realm was exactly the opposite. It would appear from some readings of contemporary sources that every house in England owned an aquarium. The *South Australian Register* (22 December 1861) made a bold claim to this effect:

> In England scarcely a drawing-room is now found without an aquarium, which is a vase or tank of water filled with living plants and aquatic animals.

Note that as late as 1861 when Australia had yet to experience its own aquarium craze, a newspaper felt it necessary to explain what an aquarium was to its readers. Although in New South Wales a few years earlier the *Sydney Morning Herald* (3 August 1857) felt no such need:

> The future historian of Great Britain will doubtless relate, among the fashions of the nineteenth century, the rise and progress of aquariums, and how ladies, grown weary of buying and losing and rebuying their cats and dogs drowned their sorrows in salt water, and transferred their affections to the lively shrimp.

---

2   For the fern craze, see Allen, D.E., *The Victorian Fern Craze* (London: Hutchinson, 1969) and Whittingham, S., *Fern Fever* (London: Frances Lincoln, 2009).

3 The domestic aquarium

Even at this relatively early date the writer is associating aquarium keeping with women and, as we shall see, this is a tenacious association which has some basis in historical reality. In the same year, the *Hobarton Mercury* (9 January 1857) reviewed an extensive article on "The Aquarium Mania" which had appeared in the English magazine *Titan*:

> "The Aquarium Mania" (would not "Wisdom" be a better word?) is a delightful paper; going into the very heart of the matter, and teaching the reader a new source of delightful amusement and instruction in a gallon or two of sea water in a tank: there is the sea and the wonders thereof in your study.

Punch (21 January 1860) saw the funny side of the aquarium craze but there is more than a little truth in a light observation in this piece, "The Advantages of Wet Weather in the Country":

> One learns to feed the parrot, and the bullfinches, and the lap-dog, and is entrusted with the keep of the vivarium which none but female hands before have been allowed to touch.

The home aquarium took the new visual experience created by the Fish House and added to the excitement of seeing fish the complex challenge of maintaining a piece of the natural world within the home:

> An aquarium is a small world ... and specifically a metonymy for the tide-pool ecologies that would claim such broad interest in England in the 1850s.[3]

This may be true, but while the aquarium metonymises a specific environment and the experiences that go with it, I think it also stands as a metaphor in which the ownership and management of a small piece of the sea reminds the enthusiast of his or her specific status as a human and, furthermore, as a human created with the duties to nature that accompany his or her God-given superiority over all other species. The

---

3   King, A., *The Divine in the Commonplace: Reverent Natural History and the Novel in Britain* (Cambridge: Cambridge University Press, 2019).

home aquarium constitutes a complex of meanings that often seem to play against each other. Silvia Granata writes of:

> the tension between the aquarium's status as commodity and its ambiguous nature as an object whose purpose was to contain living (and potentially fragile) beings.[4]

And the problems of aquarium keeping present themselves not merely as technical challenges to be faced but as responsibilities to be shouldered as the creatures so enthusiastically harvested to stock these little parlour oceans or study rockpools died in alarming numbers and all too easily.

Time and again we find references that suggest that the home aquarium was all but universal. This chapter will explore that claim. It will also explore a history by which fish and other marine creatures became visible in another new way. As Silvia Granata put it with reference to the discovery of the seaside not only as a venue for leisure but also as a site of amateur science and collecting:

> instead of simply bringing back inanimate objects, Victorian tourists could now reconstruct a living miniature sea in their own home.[5]

Christian Reiss points out that the history of aquariums:

> is not just exclusively a history of science, but also a history of fanciers of animals, of industrial and urban technology and of nature constructed. It is a history of a technology that developed from a broad desire to bring nature into domestic spaces and – last but not least – into scientific laboratories.[6]

---

4   Granata, S., "The Dark Side of the Tank", in H. Kingstone and K. Lister (eds), *Paraphernalia! Victorian Objects* (London: Routledge, 2018), pp. 48–58, p. 48.
5   Granata, S., "Let Us Hasten to the Beach: Victorian Tourism and Seaside Collecting", *Lit: Literature, Interpretation, Theory*, 27 (2016), pp. 91–110, p. 91.
6   Reiss, C., "Gateway, Instrument, Environment: The Aquarium as a Hybrid between Animal Fancying and Experimental Zoology", *International Journal*

3 The domestic aquarium

And just as Philip Gosse had advised the Zoological Society on the development and stocking of the Fish House so it was his books about the seaside and, especially, rockpools – as well as on the aquarium – that inspired a generation to take and keep all kinds of sea creatures:

> He [Gosse] helped make the aquarium a common sight in the middle class drawing rooms.[7]

Pteridomania caused lasting damage to the fern ecology of England and, likewise, Gosse's influence led to the widespread destruction of fragile marine environments, which were so denuded that they have yet to recover and probably never will. This was an issue not lost on Gosse himself and, as we will see in a subsequent chapter, it caused him bitter regret.

There is ongoing debate, as we have seen, about who actually invented the aquarium. A strong contender is the chemist Robert Warrington:

> Between May 1849 and March 1850 Robert Warrington invented the balanced aquarium.[8]

Another is Anna Thynne, whose activities with her madrepores established the possibility of keeping sea water fresh over long periods. George Johnston had had success with sponges and corals and had suggested that a marine aquarium would be viable. The French marine biologist Jeanne Villepreux-Power also has a claim and she was certainly doing experiments with sea water and sustaining an aquarium-based marine environment as early as 1832. Her claim is backed by no less a figure than Richard Owen who referred to her as "the mother of aquariophily". An anonymous author of a piece in the

---

    of History and Ethics of Natural Sciences, Technology and Medicine, 20 (2012), pp. 309–36, p. 312.
7   Smith, J., *Charles Darwin and Victorian Visual Culture* (Cambridge: Cambridge University Press, 2006), p. 60.
8   Hamlin, C., "Robert Warrington and the Moral Economy of the Aquarium", *Journal of the History of Biology*, 19 (1986), pp. 131–53, p. 131.

North American Review made a patriotic case for William Stimpson, the curator at the Smithsonian, who (without knowing of Johnston's tanks) had:

> as early as the year 1849, made seven or eight small aquaria which were perfectly successful ... To him may safely be assigned the credit of having made the first systematic attempt at constructing an aqua-vivarium.[9]

Lastly, we might also consider the claim of Nathaniel Ward. His invention of what became known as the Wardian case, a closed environment in which natural condensation kept plants and even insects alive over very long periods, transformed the transportation of botanical specimens. The case made it possible to send living specimens on long sea journeys (on deck they got plenty of light but were protected from damaging salt water) but it also made it possible to bring new kinds of plants, especially ferns, into the home and it was a short step from the Wardian case to the aquarium.[10] It has been well said that:

> The inception of the Victorian aquarium owes as much to the development of the terrarium as to the activities of marine naturalists.[11]

Indeed, the Wardian case is a foundational object for two Victorian crazes – home aquariums and fern collecting – and the case was itself a product of a specific technological moment:

> There is no question but that the availability of inexpensive window glass was one of the main factors for the rise of fern cases, aquaria, terraria and vivaria in Europe.[12]

---

9   Anon., "The Aquarium", *North American Review*, 87 (1858), pp. 143–57.
10 Keogh, L., *The Wardian Case* (Kew: Kew Publishing, 2020). This innovative study of Wardian cases does not consider their influence on aquariums.
11 Murphy, J.B., *Herpetological History of the Zoo and Aquarium World* (Malabar: Krieger Publishing Company, 2007), p. 134.

3 The domestic aquarium

But whoever you consider was the inventor of the modern aquarium, what came together by about 1850 was a method of preserving living things using a glass case, an understanding of the chemical composition of sea water and various experiments in aeration that enabled the creation of sustainable environments where marine animals could live healthily and "naturally" in captivity. It was this set of discoveries that made the Fish House possible and then a myriad of home enterprises ranging from elaborate and ornate cases, which were significant pieces of furniture in their own right, through to more modest tanks devoted to the amateur study of the sea to simple glass globes, bowls or even jam jars placed on a windowsill. And, as with all discoveries, for each new breakthrough came a new question and a new problem to solve. The Council of the Zoological Society was clearly over-optimistic in its assessment of how readily a series of large-scale tanks could be maintained and the society's problems were replicated and often not solved in many homes in England.

If you wanted an aquarium, there were several ways you could go about it. The *Ladies' Treasury* (1 May 1859) reminded its readers that: "The aquarium is rapidly becoming one of the most popular of drawing room decorations", which could be achieved for "a trifling cost" and then gave its readers some practical tips on how to set one up. The *Illustrated Times* (27 September 1856) suggested that a good aquarium could be had for 25 shillings "from any of the numerous makers who have suddenly and almost simultaneously started into commercial existence". This newspaper recommended Smith of St John Street or Bohn of Essex Street but there were many others to choose from and, if £1 5s was too much, you could always buy a pastry cook's glass shade which was "a cheaper substitute for such a reservoir". But there was a difference in these two approaches. People had kept fish in bowls long before the aquarium was invented; what we are seeing in the aquarium craze is a very precise moment of technological and

---

12   Rehbock, P.F., "The Victorian Aquarium in Ecological and Social Perspective", in M. Sears and D. Merriman (eds), *Oceanography: The Past* (New York: Springer, 1980), pp. 522–39, p. 526. See also Murphy, J.B. and McCloud, K., "The Evolution of Keeping Captive Amphibians and Reptiles", *Herpetological Review*, 41 (2010), pp. 134–42, p. 134.

commercial transition where a new kind of commodity reliant on new materials and design replaced the traditional artefact and brought with it a new way of seeing sea creatures and new ways of thinking about them.

As is often the case, a craze brought a whole raft of new business opportunities:

> Many wealthy Britons also bought or built aquariums for their houses. By 1856, the "mania" for freshwater and marine aquaria was raging at "fever point" – many had seen the financial opportunities of becoming aquarium suppliers. Prices fell as competition increased allowing ordinary Britons to also purchase aquariums for themselves. Throughout the 1850s, to use Matthew Goodman's phrase, "ponds and tidal pools were uprooted and brought indoors."[13]

But, as Lloyd put it, the craze, although well served with commodities, was not necessarily equally supported with knowledge:

> to increase the desire thus formed for having domestic fresh water and marine aquaria everywhere. And to supply it, books were written now considered of not much value, tanks were constructed in a manner which is at present regarded as of no great worth, as being utterly wrong in principle; and shops were opened to supply the rage thus suddenly awakened.
>
> (*Scientific American Supplement*, 16 August 1879)

There were numerous aquarium makers and aquarium suppliers in London in the 1850s with the main cluster being around Covent Garden. Here, in the conservatory above the piazza, it was possible, as the author of "The Aquarium Mania" pointed out, to buy a basic tank or globe, either empty or fully stocked, as well as specific fish or marine creatures:

---

13   Elwick, J., *Styles of Reasoning in the British Life Sciences: Shared Assumptions 1820–1858* (London: Routledge, 2015), p. 124.

3 The domestic aquarium

Here he would have a globe of pugnacious sticklebacks, there a tank with a glistening shoal of goldfish, with gudgeon and minnows, next to that a dish of pond snails.

The dealers in this class – the main one was called Kennedy who was sufficiently well known for the correct name for this area, the Bedford Conservatory, to be occasionally replaced by "Kennedy's Conservatory" – were clearly catering to the lower end of the market. But even that required some investment on the part of the would-be home aquarium keeper. This area was one of the sights of London and continued to trade well after the aquarium mania had passed. As late as 1913 it was described as:

> the miniature zoo known as the Bedford Conservatories where goldfish, snakes and other amphibians are sold and which is familiar to every visitor to the market.[14]

Gustave Loisel included the Covent Garden aquarium (and the aviary where birds were traded) in his survey of nineteenth- and early twentieth-century zoological displays, which confirms the relative importance of the shops and dealers in the Conservatory.[15]

However, between the refinement of the best quality shops and the rough and tumble of the Bedford Conservatories, there were other outlets. One is described in "The Aquarium Mania". The shop window, the counter and every spare inch of space is crammed with crowded tanks containing fish and other aquatic animals:

> so densely crowded together, as to make it doubtful whether the proprietor might not be amenable to the law for the prevention of cruelty to animals.

---

14 Jacobs, R., *Covent Garden: Its Romance and History* (London: Simpkin & Co. Ltd, 1913), p. 233.
15 Loisel, G.A.A., *Histoire des Ménageries*, 3 vols (Paris: Octave Doin et Fils, 1912), vol. 3, p. 315.

Except the law for the prevention of cruelty to animals didn't apply to fish.

In "My Aquarium", a paper given in Ottawa in 1893, H. Beaumont Small (who had successfully maintained a balanced aquarium at home for 30 years) looked back on the beginnings of the aquarium craze and observed that:

> In that year [1857] one of the quarterly reviews remarked that the making and stocking of these [aquariums] had created a new and important branch in commercial industry.

And that:

> In 1856 Barnum introduced into New York the first of what he styled "Ocean and River Gardens" and a few months afterwards they were for sale in all shapes and sizes for private use. Before that, the glass globe for gold-fish was the only representative of the new apparatus.[16]

In the *Boy's Own Magazine* (no date but mid-1850s), Charles Kingsley advised that a cylindrical glass jar could be had for 3 to 4 shillings and would make a serviceable, if basic, aquarium.

The craze came slightly later to France, with the display of fish at the Paris Exhibition and the opening of the aquarium in the Bois du Boulogne being especially influential, although the home aquarium had established itself in France before either of these events. The *Tasmanian Morning Herald* (19 November 1866) observed that:

> Aquariums are now the fashion, and are to be found on a small scale in almost every private house ... M. Duval, the well-known Parisian butcher [!] ... has just opened an immense establishment on the Boulevard Italians [sic] for the supply of aquariums and in which specimens of various kinds of fish will always be kept on view.

---

16 Small, H.B., "My Aquarium", *The Ottawa Naturalist*, VII (1893), p. 2.

3 The domestic aquarium

Camille Lorenzi's study of the craze in France also shows that, as in England, the home aquarium was especially associated with women, describing it as "un objet decorative pour les jeunes filles de bonne famille" ("a decorative object for young ladies of good family"). However, in France it appears that the emphasis was far less on science than in England and far more on the decorative:

> Entre les mains des maîtresses de maison des années 1860, l'aquarium n'était plus un appareil de laboratoire, mais un pur bibelot des intérieures distingués. L'aquarium était meme souvent perçu comme un tableau, un composition de forms et de couleurs, toujours recréée par les mouvements des plantes et des animaux aquatiques dans le reservoir.
>
> [In the hands of the mistresses of the house during the 1860s the aquarium was not at all a piece of laboratory apparatus but purely a bibelot for distinguished interiors. The aquarium was often even seen as a picture, a composition of forms constantly re-created by the movements of the plants and aquatic animals in the tank.][17]

This is not surprising as in France there were, at this time, no popular guides to aquarium keeping or marine science of the kind that were plentiful in England. This was pointed out by William Lloyd in his description of the Bois de Boulogne aquarium in the 1862 issue of the *Bulletin de la Société zoologique d'acclimatation*. Lloyd added that such guides would surely follow, as indeed they did, with Pierre Carbonnier's *Guide Practique du Pisciculteur* being one of the first. Like Lloyd, Carbonnier was not only a marine scientist in his own right but also the proprietor of an aquarium shop so the disinterest of science ran parallel to the self-interest of business. The French may also have faced a difficulty that the English didn't, in that measures to protect the French salt manufacturers included making sea water a contraband item and even filling a globe from the sea and bringing it indoors

---

17 Lorenzi, C., "L'engouement pour l'Aquarium en France (1855–1870)" (The aquarium craze in France, 1855–1870), *Sociétés et Représentations*, 28 (2009), pp. 253–71, p. 260.

could expose you to legal sanction. John Harper mentioned a lady who wanted to do the right thing so showed the local inspector of customs her aquarium – which he thought was beautiful. The result was that he gave her a licence to import sea water, which was just as stringently enforceable as a licence to import brandy. Notice that in the late 1850s it appears that the French customs officer appears to have seen an aquarium before. This certainly would not have been the case for the equivalent official in England.[18]

In his exhaustive study of the London working class, Henry Mayhew came across traders in fish and, specifically, goldfish. He divided them into two groups: "street fish and shellfish sellers" and "street sellers of gold and silver fish". Of this latter group, Mayhew identified about 70 in all who acquired their stock from three wholesale dealers in gold and silver fish. These might be imported from France, Holland, Belgium or further afield but about 75 per cent were bred in England where they were reared in a kind of fish farm in Essex provided with steam-heated ponds.[19]

One such wholesale company was that of Luigi Cura. He was recorded in the 1891 census as an ice-cream maker but in 1901 as a gold and silver fish importer. *The Falkirk Herald* (28 April 1923) tells us that Luigi's brother was given some goldfish in 1860 and Luigi – who was at that time working as a door-to-door seller of religious images – sold them and set up a sideline as a street seller of goldfish. By 1923 the company was importing over half a million goldfish per year and was one of the biggest companies involved in the sale of fish and reptiles. Pleasingly, the company (L. Cura & Sons) is still operating today as a fish hatchery so that present of goldfish created a business that has lasted for over 160 years and represents, perhaps, the last echo of the aquarium craze. In April 1893 John Hamlyn, the most famous London animal dealer after the Jamrach family, imported 1,200 goldfish.[20]

---

18 Harper, J., *Glimpses of Ocean Life* (London: T. Nelson & Sons, 1860), p. 361.
19 Mayhew, H., *London Labour and the London Poor*, 3 volumes (London: Griffin, Bohn & Co., 1861–62), p. 147.
20 Cowie, H.L., *Victims of Fashion* (Cambridge: Cambridge University Press, 2022), p. 206.

3 The domestic aquarium

The dealers interviewed by Mayhew had two main theatres of operation. They sold on the London streets but they did their best business in the summer months going door to door in suburbs and nearby towns where there were "villa residences of the wealthy". They bought their fish at between 5s and 18s per dozen, depending on size and quality, and sold them at 2s per pair with another 2s charged for the globe to keep them in (this was usually filled with river or rain water or even pump water and there was no aeration). Mayhew calculated that, in this way, the dealers sold about 131,040 fish per annum. One told him that:

> My customers are ladies and gentlemen, but I also have sold to shopkeepers, such as buttermen that often show goldfish and flowers in their shops.[21]

So even though Mayhew's interviews appear to reveal activity at the lower end of the trade, the evidence suggests that the market for aquariums was populated mainly by the middle classes and upwards where clergymen earned an average of £267 per year, doctors £200 and barristers £1,387.[22] These are the "ladies and gentlemen" of Mayhew's informant and we may compare their income with that of a labourer on £44 or a housemaid on £12.

To explore the higher reaches of the trade we can do no better than to pay a visit to Thomas Hall of City Road or better still to William Lloyd's shop "selling everything related to aquaria" in Clerkenwell if we are among the early adopters of the fish-keeping craze in 1855. But if we haven't caught the bug quite yet we can go to Lloyd's "aquarium warehouse" in the highly salubrious Portland Road near Regent's Park a couple of years later. His road to marine science was unlike that of many of his middle-class customers in that he had been working for a bookseller until 1854 when Gosse's *The Aquarium* – which he could not

---

21 Mayhew, H., *London Labour and the London Poor*, 3 volumes (London: Griffin, Bohn & Co., 1861–62), p. 147.
22 For a detailed account of the wages paid for various Victorian occupations, incomes and proportions of the population in different economic brackets, see https://logicmgmt.com/1876/living/occupations.htm

afford but read and committed to memory at one sitting in a shop – and his epiphany outside the Fish House led him to set up a tank of his own and from there to develop an extensive business. For 1s you could buy Lloyd's catalogue (*A List with Descriptions, Illustrations and prices of whatever relates to Aquaria*) and this enables us to calculate what the cost of setting up a home aquarium really was.[23]

First you would need a tank and a stand to put it on. A 10-gallon slope-back tank would cost you £4 5s and the stand (and I am here assuming that you want a plain stand made in good unpretentious English oak rather than flashy mahogany) £3. You could pay up to £21 at Lloyd's and all his tanks were made by Sanders & Woollcott, the company that Mitchell had employed to fit out the Fish House, which was recommended by Gosse and for which he was the sole agent. Their tanks were of glass, slate and wood, but if you had very deep pockets you could buy an all-glass tank from Lloyd & Summerfield in Birmingham. These had started life as aquariums in the old sense and were displayed as vessels for growing aquatic plants in the Birmingham Court of the Crystal Palace in 1851. However, they were now being repurposed for fish keeping in wealthier homes. Other London makers included Treggon & Co. and Phillips & Co. but, in the opinion of the author of a review of Hibberd's influential *Rustic Adornments for Homes of Taste, and Recreations for Town Folk in the Study and Imitation of Nature*, Greef & Co. in Cambridge "turn out a better and cheaper article than the London tradesmen".[24] There was also Edwards. In 1871 the amateur naturalist – what we would now call a citizen scientist – Helen

---

23 Lloyd, W.A., *A List with Descriptions, Illustrations, and Prices of Whatever Relates to Aquaria* (London: W. Alford Lloyd, 1858).

24 Hibberd was also the author of *The Book of the Aquarium and Water Cabinet* (London: Groombridge & Sons, 1856), one of the many popular manuals designed to help would-be home aquarists get started and keep going. Hibberd is largely forgotten now (except in specialist books like this) but anyone wishing to understand Victorian taste, especially in the large group of the newly prosperous but not wealthy, should read his work. A rare study is Wilkinson, A., "The Preternatural Gardener: The Life of Shirley James Hibberd (1825–1890)", *Garden History*, 26 (1998), pp. 153–75, followed up by Wilkinson, A., *Shirley Hibberd: The Father of Amateur Gardening* (Bromsgrove: Cortex Design, 2012).

3 The domestic aquarium

Watney bought an aquarium for her house in Beaumaris from his shop in Menai Bridge (she had previously kept aquariums in London and Hampshire and I believe that she was one of the people whom Lloyd set up from his shop) and noted that he was:

> The gentleman who supplied Mr Alford Lloyd with all those – or at any rate – a great many of them – nice slate tanks he had in his establishment in town a few years ago.[25]

You now need to put water in your tank and the most reliable way of doing that is to buy Lloyd's own sea water at 7s for 12 gallons (this may well have been water which Lloyd had bought cheap from the Fish House and nursed back to health) but if you were worried that you needed some back-up you could buy 6 quarts of Hain's Formula for making sea water for 12s. Gosse's recipe was to add 3.5 ounces of table salt, a quarter of an ounce of Epsom salts, 200 grains of magnesium and 40 grains of potassium to a gallon of water.

You now need to landscape your tank and for this you need a canister of cement (6d), two bags of sand (2s) and a bag of shingle (1s).

But before you can even think of putting any animals in the tank, you have to think about how you are going to manage them and how you are going to ensure that you keep the water they live in properly healthy. A good start is a dozen fronds of seaweed at 6d each (6s) to help with aeration. You now need some tools: a gutta percha siphon (3s 6d), glass tubes ("for removing offensive matter") 5s for 6, a dipping tube 1s 6d, straight-ended wooden forceps 2s 6d, spoon-ended wooden forceps 3s 6d, a hand net 9d, a glass rod 6d, a syringe 2s 6d, a specific gravity ball 1s 3d, a hydrometer 6s, a hygrometer 7s 6d, a floating thermometer 5s, and a double self-regulating thermometer £1. You also need a pump 7s 6d and, to ensure that your fish always have something to eat, a box of dried food at 6d.

You are now ready to stock your tank. Lloyd could sell you pretty much anything – in addition to numerous amateur freshwater collectors he had 14 full-time collectors of sea creatures and plants dotted around the coast – but I am assuming you are making a modest

---

25 Watney, H., "Marine Aquaria", *Hardwicke's Science Gossip*, 7 (1871), pp. 196–98.

start with six sea anemones at 12s, a dozen prawns (3s), six tube worms (6s) and six starfish (6s). This is a modest collection, but it does represent quite well the largely fish-free home aquariums of the 1850s.

All this has cost you £13 6d or slightly less than a month's pay for a doctor.

But now you want to study your animals and set up a proper marine science facility in your home. The first thing you need is an aquarium table to keep your scientific instruments and specimens on. A medium quality one will cost you £3 15s. A specialised marine microscope will cost £6 16s, object glasses will be another £2 10s, a condenser will be £1 and a lamp to illuminate everything and, especially, to optimise the performance of the microscope, will cost 7s 6d. You'll need a further support for the microscope (2s 6d), some small mirrors (6s for 16) and some small glasses (2s for 10), a gutta percha cover to protect the microscope when not in use (12s 6d) and some blue screening tissue (12s :one gross @ 1d per sheet).

Your home laboratory has cost you £16 3s 6d. And you could have spent a great deal more – this is just the basics. And certainly, aquarium keeping in many middle-class homes was accompanied by the expectation that some form of scientific enterprise would be pursued as well. It has already been noted that some tanks in the Fish House had microscopes attached and the idea that a home aquarium without a microscope was somehow incomplete may be seen from this advertisement for Mr Warrington's Portable Aquarium Microscope:

> The aquarium has now become popular and thrives in the houses of many who do not pretend to cultivate science; but how often do we find that it is unaccompanied with any adjunct for assisting the eye to scrutinize the form and structure of its living tenants.

(*Quarterly Journal of Microscopical Science*, 27, 1850)

Jabez Hogg was especially positive about this microscope:

> Mr Warrington has had constructed a very portable and economical microscope, adapted either for the examination of objects in a vivarium, or for dissecting purposes. It is packed in a

3 The domestic aquarium

neat case, and being of light weight can be carried in a coat pocket; the cost complete is £2 10s.[26]

You might want this as well as your desk microscope and it appears to have been the best of the "sea-side microscopes" that were being produced at the time. The example I have seen in the collection of the Science Museum in London is neat and of obvious quality and the wooden case could be used as the stand for extra utility.

The theme of domestic marine science and home microscopy will be discussed more in the next chapter but the association of microscopes and aquariums reached an end point in the establishment of the Mikroskopische Aquarium (Microscopic Aquarium) in Berlin. This short-lived enterprise was described in the magazine *Illustrierte Zeitung* (10 February 1877) and consisted of a room with 50 microscopes set up on tables. Visitors could go from instrument to instrument looking at the different specimens and it is surely significant that its proprietors chose to call it an aquarium, presumably because they felt that a natural connection between microscopes and aquariums existed in the public mind.[27]

Finally, you need to equip yourself for field trips both to collect more specimens to populate your tanks and to observe the behaviour and forms of specimens you already have in nature. If you are a follower of Gosse or Kingsley and have started to see nature through their eyes, you will know that real science is about encountering Creation in the open air and that keeping a home aquarium is really, as Graeme Gooday said, "a domestic extension of field work".[28] You will need: a net 8s 6d, a geological hammer 7s, two stone wicker covered bottles for collecting sea water (13s), a tin box (8d), an animalcule collection bottle (7s 6d), a pocket magnifier (12s 6d), and a large hand glass (£1).

---

26 Hogg, J., *The Microscope: Its History, Construction and Application* (London: Herbert Ingram & Co., 2nd edn., 1856), p. 51.
27 Hausman, K. and Machemer, H., "The Microcosm under the Microscope: A Passion of Amateurs and Experts", *Denisia*, 41 (2018), pp. 1–46, p. 12.
28 Gooday, G., "'Nature' in the Laboratory: Domestication and Discipline with the Microscope in Victorian Life Science", *British Journal for the History of Science*, 24 (1991), pp. 307–41, p. 312.

Your field-work kit has cost you £3 9s 2d. I am assuming that you already have things like rucksacks, creels and waders.

So, a well-set aquarium together with the equipment to conduct the kind of marine science that offered the instructive element of this delightful pastime would come in at £34 6s 4d – £36 16s 4d if you decided to buy a Warrington's microscope as well. My contention is that this suggests that aquarium keeping in the style represented by the many aquarium books (and I have not included the cost of your marine science library, journals and aquarium-keeping manuals in this total) was far from the almost universal hobby portrayed in the newspaper and magazine accounts I have cited and, of course, many others.

Let's now explore who could have afforded this sort of outlay. Lloyd himself could not have done. His original interest in fish was thwarted as he could not even afford entry to the Fish House when it opened and only managed to see it after he wrote to Richard Owen asking for a free ticket. Owen obliged, Lloyd was able to see the new facility, and from this visit was inspired to decide to give up his work at the bookshop and set up exclusively in the aquarium and fish business. A rather sad picture is of Lloyd first seeing the Fish House "from its outside, through its glass walls, on Thursday, November 18th, 1852, at noon" (*Scientific American Supplement*, 16 August 1879). This was, of course, before the facility was even completed. And that precise memory of the day and time bespeaks Lloyd's infatuation with the idea of fish and aquariums at that early date. When he got home, he found the money to set up a small globe with waterweed and populated it with minnows. This was his first attempt to keep a freshwater aquarium and, from this humble start, grew the aquarium mania.

Let us take two eminent Victorians who both had incomes of £500 per annum at about the time of the home aquarium craze. One is Florence Nightingale who was set up by her father as an independent woman with an annual income of £500 (which was subsequently increased) and the poet Arthur Hugh Clough, who was set by his would-be father-in-law the task of finding a job paying a minimum of £500 per year before he would be allowed to marry his daughter. After a two-year search, he found a suitably remunerated post as an examiner in the Education Office in Downing Street.[29] An annual salary of £500 certainly put you solidly into the ranks of the middle class but to spend

3 The domestic aquarium

nearly 7 per cent of this on setting up a home aquarium would still require some consideration.

If we set an income of roughly £500 per year as a benchmark for affordable aquarium ownership, how many of the population would this have included? In 1861 (by which time the home aquarium craze was waning but which is proximate enough to offer a reasonable guide to broad numbers) 49,500 people had incomes of over £1,000, which put them into the upper class, while 150,000 were in the £300–£500 bracket, which put them at the top of the middle class.

Although £300 per annum would mean that a well-set-up aquarium would represent a major purchase, it would nonetheless be possible to indulge in the hobby at some level given that kind of income. So, 199,500 people had incomes sufficient to buy an aquarium without too much stress; and this represents 0.8 per cent of the total population (in 1861) of 23,085,579. This figure of course represents only male wage earners or women of independent means and not the total number in their household. If we arbitrarily assume a household number of six (i.e., a male head of household and five dependants) – and this is probably a slight overestimate for middle-class and upper-class families where fertility rates were, on average, lower than for the working classes – we arrive at a figure of 4.2 per cent of the population who may have had access to a high-quality aquarium in their own home distributed between less than one per cent of total households.[30] This is far from the picture of an aquarium in every house that is suggested by contemporary journalism. We may compare this figure with those of another craze, "hippomania". In 1850, the arrival of Obaysch the hippopotamus at London Zoo created a sensation and 227,938 visitors went to Regent's Park just to see him. This figure compares reasonably well with the figures for possible aquarium owners, especially as seeing the hippo cost only 1s (which was still a lot of money for the majority of the population – as we remember from Lloyd's inability to pay to go into the Fish House). The population of London in 1861 was 2,803,034

---

29 Bostridge, M., *Florence Nightingale* (London: Penguin, 2009), p. 161.
30 Pooley, S., "Parenthood, Child-rearing and Fertility in England 1850–1914", *History of the Family*, 18 (2013), pp. 83–106.

(an increase of 500,000 on the population of the city that first hosted Obaysch) or just under 10 per cent of the national population.

From these figures it is probably also safe to assume that most people who could afford an aquarium on a serious scale were based in London. This would not be exclusively the case and the fact that Lloyd's shop had arrangements for sending things by mail to anywhere in the country demonstrates that there was a market in the country and regional cities for his goods. But the simple facts of the distribution of people and income in Britain during the 1850s means that most of the aquariums which were supposed to be "in every house" were in London. Lloyd himself said that he had a list of 8,000 people who had indicated to him that they were interested in aquariums (*Hardwick's Science Gossip*, 1 July 1865) and this is a good indicator of the market for aquariums of the very best kind.

We might also compare the costs of goods and services that might have competed with aquariums for a slice of a middle- or upper-class income. At the top tier, one could pay between £4,000 and £8,000 to lease a box at Her Majesty's Opera House or £25 for 20 minutes in the company of a high-class prostitute such as Skittles. Perhaps a more telling comparison is the £60 that Jane Carlyle spent on a Clarence (a light carriage) in 1864. This was at a time when her husband Thomas Carlyle's income was probably averaging over £1,000 per year and was enhanced by a legacy of £2,000 he had just received. The Carlyles could certainly have afforded an aquarium.

Even so, it appears that as the 1850s progressed, the popularity of aquariums made them more accessible to less affluent citizens. This would include those who could only afford 4s for a pair of goldfish in a bowl of course, or even just 2s for the goldfish that they then kept in a jar of some kind. A writer in *Excelsior* (2, 1856) thought that "a washing-basin, or a soup-tureen, will answer admirably; or a delft foot-bath, or a milk-dish or a brown earthen pan", while the reader of *Household Words* (9, 1854) was told that "an aquarium may be made in a doctor's bottle or a pudding basin".

But between the people who could buy a top-class set-up from Lloyd and those keeping goldfish in a basin there were many who could afford a slightly more substantial arrangement. *The Illustrated Times* had suggested 25s for a decent tank and if we return to Lloyd's we

3 The domestic aquarium

can see that the seaweed necessary to maintain balance and a small selection of marine animals would cost a pound or so. But it may have been cheaper to buy them from the various dealers in Covent Garden or from Thomas Hall in City Road who had transitioned his business from dealing in shells, minerals and fossils in 1849 to "marine and fresh-water stock" as the aquarium craze took hold.[31] Assuming that you would make do with an existing table, windowsill or sideboard, cobbled together your tools and didn't keep an obsessive watch on water temperature etc., you could set up a reasonable tank for a little over £2. You could also bring everyday items into play rather than buying specialised tools. For example, W.E. Simmons, writing in the *Popular Science Monthly* (4, 1874), recommended that glove stretchers could be used instead of tongs to pull dead fish or rotting plants out of the tank. In the *Home Friend* (1856), the author of an article entitled "The Aquarium Simplified" thought that an economic aquarium could be set up using a 10s propagating glass holding 8.5 gallons of water, a basic stand (1s 6d), 6d worth of Portland cement, 3s worth of waterweeds and four dozen snails (important for keeping the tank clean and costing about 2s 6d), and a circular glass cover (about 1s from Claudet & Houghton) completes the set-up. Then let's assume that you simply wish to keep a pair of goldfish (4s) – that all comes to £1 2s 6d. All this brings a considerably larger proportion of the population into play although it still represents a week's wages or more for the great majority of the unskilled working class or many domestic servants.

John Harper, writing in 1858, noted that a tank could cost anything between 3 or 4 shillings and £20. Harper's own tanks were cylinders costing 4s each and with a disc of glass placed on top as a lid. He claimed that all that was needful then was a glass syringe, a pair of gutta percha tongs, a camel-hair pencil and an ivory crochet pin:

> The cost of which is so trifling that the poorest person might manage to procure them.[32]

---

31   McKie, E., *Thomas Hall of City Road: The Family in a Museum* (London: CreateSpace Independent Books, 2013).
32   Harper, J., *The Seaside and Aquarium* (Edinburgh: William P. Nimmo, 1858), p. 170.

These might be the basic necessities but Harper then goes on to describe how to solve various problems in the aquarium by using other implements that aren't mentioned here. An ivory crochet hook doesn't sound cheap either. So, although it might just be possible to set up a small aquarium on Harper's budget, it would probably be very difficult to maintain it.

Shirley Hibberd (who kept both a freshwater and sea-water aquarium over several years) who, after Lloyd and Gosse, was probably the most influential writer on aquariums (although the instructions he gave, especially regarding the number of fish one might keep in a single tank, would be unlikely to lead to success) recommended a Sanders & Woolcott tank for £5 in his *Book of the Aquarium* and, in his *Rustic Adornments*, a very slightly smaller one by Treggon & Co for £3. This is an interesting example of early sponsorship and product placement. It is notable that Treggon & Co were, in 1843, trading as plumbers and zinc manufacturers and had clearly realised that their skills in metal work and waterproofing were transferable to the newly lucrative trade in home aquariums. The review of *Rustic Adornments* in the *Home Friend*, which was quoted above, recommended a much smaller collection than that proposed by Hibberd and suggested three small goldfish (6d each in Covent Garden), two brown and two Prussian carp, also from Covent Garden at a similar price, and then a selection of native freshwater fish, which could be netted with a little diligence: six sticklebacks, two perch, four minnows, two bleak, two eels and the usual complement of water snails. He also recommended that, while most of the plants you needed for aeration could be found for free in ponds and ditches, it was worth buying some *Valisneria*. This could be expensive – 1s 6d per root in the Arcade at Brighton Station, 1s in a shop in King William Street – but could be tracked down for 3d in Covent Garden. If we add the usual Portland cement, we find that a decent tank can be acquired and stocked for about £3 8s. This represents an outlay readily achievable for people in the upper end of the £300 to £500 income bracket and gives a sense of what the socially mobile readers of *Rustic Adornments* might have considered feasible to spend in order to set up their home aquarium and thus take part in a hobby or craze that they associated with the lifestyles to which they aspired. These are, of course, figures for a freshwater tank;

3 The domestic aquarium

a marine tank would have required expenditure at a higher level. As Anne Wilkinson said of the popularity of *Rustic Adornments*:

> It made everything accessible to the aspiring middle classes, who needed to be guided, not only just in taste, but also in the practical side of setting up aquariums and summerhouses ... without spending too much money.[33]

The idea that people in a still lower income group might have afforded a small aquarium is touted by the author of "The Aquarium Mania":

> The expense, however, was something considerable at first; and for a while it was only in ... the mansions of the well-to-do in the world, that you had any chance of encountering an aquarium. But an increase in dealers and the competition between them, have made matters different now; and the man with only a few shillings to spare may yet have his collection of pet sea anemones.

Lloyd himself, who rarely had anything to spare before he went into business, certainly did pull together a penny here and a penny there to enable him to set up a basic marine aquarium in the most remarkable way:

> So with artificial sea water prepared by a Holborn chemist [I suspect this was William Bolton who was advertising his own sea water salts as early as 1854] – which salts he kindly gave me because I gave him the recipe for mixing them – I set up small aquaria in wide-mouthed glass bottles costing a penny or two each. The sea I had never seen, and was not so presumptuous as even to hope to see it, and I knew of no one living by the sea who could send me marine animals. But that daunted me not, for I used to sally forth at dead of night where heaps of oyster shells were thrown by day from street oyster stalls in Smithfield and St John's Street and bring them home. The oysters devoured

---

33 Wilkinson, A., "The Preternatural Gardener: The Life of Shirley James Hibberd (1825–1890)", *Garden History*, 26 (1998), p. 156.

in such poor neighbourhoods were not the genteel little smooth "natives" eaten at luncheon bars, but big rough "commoners," with bold foliations of the upper shell and deeply ribbed on the lower one, and in and between these hiding-places I found many little sea anemones of several species, some hopelessly smashed, but others quite perfect (having been protected by the projections of the oyster shells) and unharmed by rain or other fresh water ... All these I used to pick off the shells with never-wearying patience and care, and drop them into the factitious sea-water, and transfer them to my bottles, to which they adhered and made themselves happy. I used to feed them with little morsels of oyster flesh which I found.[34]

There would have been others like him who scraped along on low wages and in a small family apartment, which afforded some domestic privacy, but few who pursued their interest with such determination. And what a remarkable thing it is that the man who was destined to become the most important aquarist of his age – from a historical viewpoint the most important ever – had never seen the sea. However, it would still be the case that the great majority of the working class would not have been able to keep even a basic set-up not least because their characteristically crowded and frequently insanitary living conditions would not have given them anywhere to keep it nor any access to the fresh water they needed to maintain it.

But we may now assume we can include the group living on between £100 and £300 per year. In 1861 this would add another 850,000 people to the number who could have bought an aquarium (although while £100 kept you out of poverty it didn't allow much room for bigger purchases). This is slightly under three per cent of the total population and, when added to the figures for the upper and better off middle classes, we see that the probable market for aquariums consisted of about 1,049,500 people and their households or about 3.6 per cent of the total population. If we make the same assumption about household size for this group as we made for the upper and middle classes, we arrive at a figure of 5,100,000, which added to the 1,194,000 estimate

---

34 *Parlour Aquariums*, parlouraquariums.org.uk 2018.

for upper and wealthier middle-class households means that 6,294,000 people, or about 22 per cent of the population, might have lived in houses where there were aquariums. This looks more like a craze but it throws into sharp relief the meaning of a comment in the *Ladies' Treasury* (1 May 1859) that "the aquarium is rapidly becoming one of the most popular of drawing room decorations" and that this can be had for "a trifling cost". One can see the growth in aquarium sales by looking at Claudet & Houghton's stock advertisements: in 1856 they did not mention aquariums as one of their main items of trade (and by this time they must have been selling plenty of aquarium-related products) but by 1868 and in subsequent years aquarium supplies were a prominent feature. Similarly, by 1858, *Aquarium tanks* were a new and separate subsection in the Patents' Register. The mania also spread to Germany and France and an article in *Die Gartenlaube* (33 (1854) – so only a year after the opening of the Fish House) gives a glorious picture of the joys of a home aquarium as they might have presented themselves in a comfortable middle-class household:

> Der tyrannische, allgewaltige, unbändige Ocean fluthet nun auf unserem Tische und wir können nun das Leben aus der Tiefe auf dem Tische studieren, in Schlafrock und Pantoffeln.
> [The tyrannical, all-powerful, untamed ocean now flows on our tables and we can now study life from the deep on the tables in dressing gown and slippers.]

Karl Möbius, who kept tanks in his houses in Kiel and Vienna and who went on to oversee the development of the Hamburg Aquarium, spoke of:

> the enthusiasm for aquariums sweeping in from England.[35]

However, it was probably worth the investment to buy a professionally made tank rather than attempt to make your own. As the American

---

35   See Nyhart, L.K., *Modern Nature: The Rise of the Biological Perspective in Germany* (Chicago: University of Chicago Press, 2009), p. 134.

writer W.E. Damon put it in his *Ocean Wonders: A Companion for the Seaside*:

> A home-made tank will do then? My dear reader, you cannot anticipate the difficulties; you had better attempt to build a house or a railroad with all its rolling stock, than construct an aquarium which will not leak on the parlour carpet, or kill the fishes it contains.[36]

This sounds like good advice – but it also alerts us to another feature of the aquarium craze, which is that it made the home aquarist dependent to a greater or lesser degree on a commercial professional and an industry. It simply was not possible to have an aquarium as a hobby or object of scientific interest without entering into relationships with a world of commodities (some of which were living things) and monetary transactions. In this, the aquarium craze perhaps points to a new or developing relationship between the interests that constitute any individual and the outer world of capitalism and trade in that no longer was it possible to develop a pastime purely in a sphere of domesticity or self-sufficiency. To have and develop any interest involved becoming a consumer and entering into a chain of commodity exchange that you both constituted and were constituted by. And although Damon writes of "Miss Elizabeth Emerson Damon of Windsor Vermont" as "the pioneer inductor of the private aquarium in this country [the USA]", he admits that her initial efforts were based on a 2-quart jar with weeds, snails and tadpoles.[37] Aquarium keeping of that sort would not sustain a craze.

Children were also keeping fish and, it appears, taking a precocious interest in marine science. A review of Thomas Rymer Jones' *The Aquarian Naturalist* offered the following vignette of the aquarium craze at work in an upper middle-class home:

---

36   Damon, W.E., *Ocean Wonders: A Companion for the Seaside* (New York: D. Appleton, 1879), p. 187.
37   Damon, W.E., *Ocean Wonders: A Companion for the Seaside* (New York: D. Appleton, 1879), p. 186.

3 The domestic aquarium

To very many it may seem the highest praise we can bestow on Professor Jones' work to rank it in the list of those which make the study of Marine Zoology too easy. There is some likelihood that when *Miss Mary* and *Master Thomas* have glanced over it, they may hasten with collecting jars to the shore to fill the aquarium which fond paterfamilias has set up for them; and, when it is well-stocked, the wondering parents may be called to listen to the young lady's remarks about "Sir John Dalyell's discoveries," or to the hopeful brother's correction, "You forget, sister! What you refer to is found in a remarkable paper by Quaterfages, in the *Annales des Sciences Naturelles*," or, "The observation is by Kölliker," while the accomplished Miss takes her revenge by adding, "Here is Professor Jones' *Aquarian Naturalist*, and you will find from it that we were both mistaken. He says that van Beneden first noticed it." Let not any reader smile: the picture is from life.

(*North British Review*, 32, 1860)

Less precocious children might have made do with Elizabeth Allom's *Sea-side Pleasures* which, as early as 1845, showed well-to-do children learning about marine science as part of their holiday in Ramsgate or, later, the Reverend William Houghton's *Sea-Side Walks of a Naturalist with his Children*.[38]

But while the main aquarium craze had largely died out by the early 1860s, the most frequent references to child fish keepers come in the 1870s and magazines like *Little Folks*, *Chatterbox* and the *Boys' Own Paper* among many others were carrying articles on aquariums. Lloyd contributed an article entitled "A Pet Jack" to *Chatterbox* (24 February 1874) that exhorted his readers:

> Boys, there really is no more innocent fun, more to make you laugh, and less danger in scarring your consciences and becoming

---

38 Allom, E.A., *Sea-side Pleasures* (London: Aylott & Jones, 1845); Houghton, W., *Sea-side Walks of a Naturalist with His Children* (London: Groombridge & Sons, 1870).

cruel, in my plan of keeping a Jack [a small pike] in a big tank in enjoyable captivity, than in wantonly taking his life and calling it "sport".

Lloyd was always aware of the welfare of captive fish and, arguably, the bankruptcy he suffered when his shop went out of business was due at least in part to the fact that he was more interested in the fish and marine science than in selling aquariums. But even this flourishing of aquarium keeping among boys appears to have been of flimsy substance as these advertisements show:

A small globe aquarium, a prize bow and fancy arrows, and thirty yards of net-work suitable for a fowl-house or garden fruit will be exchanged for a pair of skates.

(*Boys of England*, 25 January 1868)

AQUARIUM – 2ft 7inches long 21 in. wide and 15in. deep, enamelled ends and bottom, plate-glass front, opaque glass back, polished oak cornice. Wanted, a needle rifle. Open to offers.

(*Exchange and Mart*, 1868)

Notice that here we have unwanted aquariums at each end of the scale of expense. One is a simple goldfish bowl, the other an elaborate and expensive piece of furniture.

When an epidemic of typhoid threatened Uppingham School, the boys were evacuated to the fishing village of Borth in Ceredigion. According to *The Times* (14 April 1875), one of the first things they did was to establish an aquarium; to stock it, they went out and caught an octopus.

We also see evidence of a second wave of the aquarium hobby in magazines aimed at women. The *Ladies' Treasury* (1 September 1873) reviewed Cornell's *Popular Recreation* with special approbation for Lloyd's article on the home aquarium. We meet Lloyd again writing in the *Englishwoman's Domestic Magazine* (1 December 1877) where he recounts the story of the aquarium pioneer Anna Thynne by annotating an excerpt from her journal in the most glowing terms:

3 The domestic aquarium

"In the spring of 1847 I wished to try whether I could adjust the balance between animal and vegetable life, and sent for shells and small pieces of rock to which living seaweed was attached."

This is a remarkable sentence ... because for the very first time known to be on record in print, an intention is here expressed to make and sustain in an aquarium, though it was not then so called, for it had no name, an arrangement in which plants and animals, should in a reciprocal manner aid each other's existence, as they do in the state of nature.

Lloyd returned to the theme of practical aquarium keeping in the same magazine's issue for 1 June 1878.

My assumption is that the reason that there appears to have been a resurgence of interest in the home aquarium in the 1870s, ten years after the initial craze had faded, was the growth in public aquariums at this time and the speculative bubble which inflated beneath their construction in dozens of locations. This will be dealt with in detail in two subsequent chapters but it is worth noting now that just as the home aquarium craze grew out of the failure of the Fish House so the public aquariums grew out of the failure of the home aquariums. The problem that faced everyone who tried to keep an aquarium on whatever scale was that it is difficult to do without constant attention and, even then, it might fail. As Judith Flanders has put it:

> Tanks operated on the balance theory (the understanding that the plants would produce enough oxygen for the fish to survive). As with Wardian cases, in reality the balance theory was one of those nice ideas that did not quite work in practice – the plants in a small tank could never produce enough oxygen – in practice it was mechanical pumps which made aquariums possible for the amateur for the first time.[39]

But not everyone knew this, nor could everyone afford a pump and so fish died in massive numbers, whether these were the goldfish

---

39   Flanders, J., *The Victorian House* (London: HarperCollins Publishers, 2003), p. 165.

condemned to live just a couple of days unfed in an unaerated bowl – "did you ever keep gold-fish in a glass globe? If you have, you will be pretty sure to remember that you kept them but for a short time, for they died off rapidly; and you got tired of replacing them" (*Godey's Lady's Book*, 54, 1857) – or elaborate bouquets of sea vegetables and scuttling crustacea gently rotting in the corner of a middle-class drawing room. In *Household Words* (9, 1854), the challenges of keeping sea water fresh were made light of:

> This balance is not very difficult to get but if not it is only necessary to take every morning a portion of water out of the aquarium and allow it to drip back from some little height into the vessel.

It was this kind of underestimation of the challenges and optimism about outcomes that led to so many problems in the parlour aquarium. Throughout the 1850s, in addition to the various instruction manuals and lectures, the scientific periodicals like *The Zoologist* and the *Quarterly Journal of Microscopical Science* were running dozens of articles on aquariums and this surely reflects the need to have answers to the questions daily posed by an aquarium. A magazine like *Leisure Hour* could run a six-part series called "The Little Aquarium in the Parlour Window" detailing the successful maintenance of a balanced tank in the home. But the average home aquarium keeper seems to have had little idea of how to keep the tank healthy. An indication of how basic the need was might be indicated in a short article in *The Naturalist* (7, 1857) which dealt with "filling and emptying an aquarium". What might seem to be the most basic task was, in fact, not easy when running water from pipes was only intermittently available and less than 10 per cent of houses could afford the 3s per week bill from the water company. Yet it is something that was usually skirted over in the extensive aquarium literature. This may be suggestive of the availability of running water to the authors or it may speak to an assumption that aquarium keepers would have access to a reliable source. Which is an assumption about their income.

Opinions varied on how long it took to maintain a healthy aquarium. Lloyd said that:

3 The domestic aquarium

Five minutes of vigilant attention paid regularly, *and as a habit*, once a day in the morning is more than sufficient to keep in order any of the tanks here named; but when this duty is once well performed there should be no more fidgeting with the animals.[40]

This was for a balanced tank where the water was not changed but kept healthy through the action of aerating plants. On the other hand, Charles Harris' article "On the Marine Vivarium' argued that:

An hour or so in the early morning is all that is required.

(Annals and Magazine of Natural History, 15, 1855)

But Harris believed that keeping water fresh through the balancing method was too difficult for the parlour aquarium keeper to operate successfully, so he simply changed his water when it needed changing and kept his tanks clean and fresh at other times. Even allowing for the significant difference in Lloyd's and Harris' approaches, there is a big gap between five minutes and an hour (or so …) and the most notable feature of that gap is what it implies about the class of the aquarium keeper. The person who had to serve in a shop or push a quill in a legal office would not be likely to have an hour or so spare in the morning. Lloyd himself worked from 8 am to 9 pm in the bookshop so although someone of his dedication might have got up at 5 am to attend to his tank, few others working those crippling hours would have had the energy. Test cricket matches in England start at 11 am. The original reason for this was that this enabled professional men and men of business to get an hour in at the office before getting a cab from the city out to Lord's for the start of play and these were the men who also had time for an hour a day with the aquarium. Even more so, their wives and daughters who would not have worked and would have had light, if any, household duties could have taken their time to maintain the home aquarium. So, what we see in the varying estimates of how much time was needed to keep an aquarium healthy is not only a function

---

40 Lloyd, W.A., *A List, with Descriptions, Illustrations and Prices, of Whatever Relates to Aquaria* (London: W. Alford Lloyd, 1858), p. 144.

of different methods but also a function of expectations about class – manifested among other things by an expectation of access to clean water to enable regular replenishment of the tank instead of balancing – and it is no surprise to the see the working-class Lloyd looking to be as sparing of time and water as possible.

It was even possible to combine two different enthusiasms by buying an aquarium that also contained a bird cage – some even had a glass structure projecting up into the cage from the tank below so that the birds would have fish swimming in their midst – but even these didn't necessarily survive the waning of the craze:

> [For sale] A fine large Aquarium and stand containing birds and fish.
>
> (*The Belfast News-Letter*, 9 April 1864)[41]

From early on, aquarium keeping (and, as we shall see, conchology and phycology) were seen as especially suited to women and many illustrations of home aquariums show women tending them or watching them. The *Hobarton Mercury* (9 January 1857) summed it up well:

> Let the intending proprietor of an aquarium content himself (or herself, we ought, perhaps, to say; for aquarium keeping properly attended to is pre-eminently a lady's occupation).

The reasons for this will be explored in another chapter but we should remember that, in the first bout of the aquarium craze, the object was not only to keep an interesting selection of marine life alive in the tank but also to produce an aesthetic object for the home. Advice on the decoration of aquariums formed an important part of the early instruction books and this comprised both the addition of various ornaments to the tanks and the creation, by using different sands,

---

41 A high-quality group of this kind of aquarium is discussed in Ricke, H., "Der Vogel im Goldfischglas" (The bird in the goldfish bowl), *Journal of Glass Studies*, 59 (2017), pp. 261–84.

3 The domestic aquarium

rocks and gravels, of little underwater seascapes. These were miniature environments, perhaps copied directly from the colour plates in one of Gosse's books. An Australian example was observed in a notice in *Bell's Life in Sydney and Sporting Review* as early as 8 May 1858:

> There is on exhibition at Mr Lea's shop in Park Street, a very interesting specimen of art. It is termed an Aquarium [note that the writer did not expect the reader to know the word], and the interior is tastefully formed in imitation of a submarine grotto … through and around which a great number of little fish glide and disport themselves.

And the aquarium aesthetic was not only about fish tanks: on Christmas Eve 1867, *The Age* reported that a Mrs Buckler went to the fancy dress ball held in honour of the visiting Prince Alfred dressed as an aquarium. And *Punch* (7 March 1868) reported on an aquarium costume spotted in Paris:

> the waist enclosed with beautiful shells, the continuation being a short jupe of sea-green colour on which are to be found numerous samples of the animal and vegetable world.

adding:

> Only imagine a man's feelings at hearing that he has to waltz with an aquarium.

The aquarium craze even penetrated the German imperial court with the *Illustrated London News* (24 March 1877) reporting that, as well as a six-foot-long sausage, the kaiser's birthday presents included:

> a giant aquarium full of jelly containing eighty fish ready for the table.

The penknife and pair of scissors also on the list must have felt like an anticlimax.

But the lighter hearted side of the aquarium craze did nothing to address the problem of dying fish. This had been encountered in the Fish House (although quite successfully concealed, it appears) and now reared its head in the home. Lloyd was very aware of the issues that keeping an aquarium (or rather failing to keep an aquarium) posed to the welfare of the fish and before he would sell you any stock, he insisted that your aquarium was properly set up. His catalogue included a questionnaire that had to be filled in by anyone wishing to buy one of his tanks and this included questions about light, placement (near a window? near a fire?), temperature control and all the fundamentals that would ensure that the home aquarium was properly sited and kept in conditions that would optimise the chances of maintaining a healthy environment for the fish. Ideally, Lloyd preferred his customers to send him "a rough plan, showing the position of windows and the direction of the sun upon them, as variously influenced". Placement was vital. As Shirley Hibberd said of his freshwater aquarium:

> The situation of this tank is the secret of its success.

Lloyd would certainly not have been keen on the expensive displays in the drawing-room windows of the wealthy and, indeed, if the experience of the Fish House is anything to go on, aquariums so placed were almost certain to fail without a huge amount of effort on the owner's part. Lloyd's concern for the welfare of the plants and animals he sold is rare for the time and, as I argued above, is still relatively rare as fish appear to occupy a different moral status from that of other animals. As Silvia Granata pointed out:

> Indeed, tank residents were considered more expendable than other domestic animals. While in some cases they are playfully called "pets" or "companions," they are frequently defined as "objects," "things" or "ornaments."[42]

---

42 Granata, S., "At Once Pet, Ornament and Subject for Dissection: The Unstable Status of Marine Animals in Victorian Aquaria", *Cahiers Victoriens et Édouardiens* (en ligne), 88 (2018).

3 The domestic aquarium

And this objectification of marine life – made easier by the still-persistent belief that fish can't feel pain – meant that the death of a goldfish after a few days unfed in an unaerated bowl or a £20 tank from Lloyd reduced to a slimy and stinking swamp in a corner was of little moment. The *Illustrated London News* (30 December 1871) pointed out that the defective design and lack of a sophisticated filtration system led to appalling mortality among the fish in the Fish House and that:

> These considerations added to the almost universal failure of domestic marine aquaria from the same causes, tend to impress the belief that success was impracticable and impossible.

In the mid-1850s this had yet to be discovered but by 1860 the craze was almost over and by 1862 it was finished:

> Soon after this period (1852) the marine aquarium became a fashionable amusement which reached a fever of excitement in 1860 – something like potichomanie or crochet work – and appears, on Mr Lloyd's authority to have nearly died out in 1862.[43]

Lloyd had good reason to think that the craze was more or less dead by 1862 as he was declared bankrupt on 21 July of that year. The great majority of aquarium owners appear to have discovered what Lloyd had tried to tell them all along, which was that keeping a marine aquarium was possible but that it required a great deal of attention and expertise to keep the plants and animals healthy. In 1860 the initial volume of the magazine *Recreative Science* published a review of Shirley Hibberd's *Management of Aquaria* that confirmed the sad state of affairs:

> The "aquarium mania" may be considered as fairly dead: it died out properly and completely ... We rarely hear of "aquarians in trouble" nowadays, because the thousands who set up aquaria, without the least idea that to be successful they must be managed

---

43 Hughes, W.R., *On the Principles and Management of the Marine Aquarium* (London: John van Voorst, 1875), p. 14. Potichomania was a craze for painting cheap glass vases to look like porcelain.

on philosophical principles, have long ago given them up as "troublesome".

One could learn from the many books available and also from the public lectures that figure prominently in newspaper advertisements at this time, but this learning had to be applied in the almost daily exercise of care and maintenance, which, as this review and many other sources make clear, was beyond the ability of most amateur aquarists.

The Ladies Cabinet of Fashion (no date but probably 1857) tells a typical story:

> We three determined on having an aquarium; it would be so pleasant to see the sticklebacks fighting as our books said they would ... [We] bought a large confectioner's glass, three walking sticks, three iron hoops, made nets and fastened them to the hoops to make fishing nets. [The women then catch a dozen sticklebacks and various other creatures and put them into their aquarium.] Next morning – crash! – every fat stickleback was on his back and everything else was dead except the next who was dying ... Again we caught a mass of fish etc; again we had an aquarium; again we went to bed; again got up; every perverse fish dead on his back as before ...We cleaned our aquarium and put it on one side.

Although this is a light-hearted narrative designed to amuse and to poke gentle fun at the aquarium craze, it is also a narrative about cruelty and neglect. And we see very readily how the fate of the sticklebacks does not arouse the same emotions as would the fate of a puppy or kitten.

The Era (14 October 1860) perhaps summed up the problem most graphically and provides an epitaph for the aquarium craze:

> We are pleading the causes of humanity in pleading for a better spirit in those who undertake the care of these poor animals. No one has a right to use them as the fashionable toy of a few weeks' amusement, and then leave them to perish by miserable torture or idle neglect. We wonder at nothing where fashion leads the

3 The domestic aquarium

way; but the delicate ladies who set up an Aquarium should bear in mind that they know very little of the life they are tampering with; and though sea anemones dying by slow degrees in a foul untended bowl, create no noisy horror like a bull-fight, there is a peculiar enormity in the silence with which these murders are done, and the indifference with which they are viewed.

An article on the "Parlour Aquarium" in *The Family Friend* (1856) referred to "the fish globe as a sort of crystalline hole of Calcutta". And a writer in *Once a Week* (1, 1859) discusses the virtues of studying hermit crabs and the enormous interest these animals offer the aquarium keeper but glumly adds "these comical little crabs do not live long in a state of confinement". In *Life beneath the Waters* (the first book to be published on aquariums in the USA), the author, A.M. Edwards, reflected sadly that:

> It is said that Madame Pompadour, to her other sins and weaknesses, added that of setting this fashion of studying, really, the tortures of poor fish.[44]

But, as late as 1888, we can still encounter a casualness towards the lives and welfare of marine creatures:

> Go out at extreme low tide and with a net scoop up almost anything you can find which has been left by the tide. Take the result home and sort it over, culling out what you want.
>
> "A Marine Aquarium", *The Decorator and Furnisher*, 11 (1888)

And, one assumes, simply leaving the rest to die. As George Eliot wrote:

---

44 Edwards, A.M., *Life beneath the Waters: or, the Aquarium in America* (New York: H. Baillière, 1858), p. 16. West, R., "Fresh and Salt Water Aquaria", *Report of the Commissioner of Agriculture for the Year 1864* (Washington: Government Printing Office, 1865), pp. 446–70 gives New York's Greenwood of Broadway as the chief supplier of aquariums in the USA, p. 449.

You would laugh to see our room decked with yellow pie-dishes, a footpan, glass jars and phials, all full of zoophytes, or molluscs, or annelids – and still more to see the eager interest with which we rush to our "preserves" in the morning to see if there has been any mortality among them in the night.[45]

Notice that while in the more developed phase of the domestic aquarium craze the sea creatures are often called pets, here Eliot chooses the word "preserves", more usually used of pickled or bottled fruit and vegetables.

At a time when the home aquarium craze had long finished (although I think that the new enthusiasm for public aquariums did stimulate a brief revival in the 1870s), a comic piece in *The Graphic* (5 November 1873) recorded, I suspect quite accurately, the fate of many home aquariums and the animals that inhabited them:

> "You don't think, George, I'm going to have that filthy tank set up in my drawing room?" is Mrs Tippety Lorrup's greeting to her poor subdued husband, who in a moment of rashness has bought an aquarium without consulting her ... So the aquarium is relegated to the window of the kitchen stair case where it soon gets a rich Weston-super-Mare-ish smell.

Whatever fish George had bought are, of course, long dead and after two attempts to sell the aquarium through the *Exchange and Mart* it is eventually sold with two cracked panels for whatever someone will offer.

When Lloyd, writing in *Hardwick's Science Gossip* (July 1865), himself looked back on the craze for aquariums in 1865 he reflected that he had set up 3,548 aquariums for his customers and, of those, he believed that only 48 were still in operation. Lloyd's business ran between August 1855 and July 1862 – a total of 79 months – which means that, during that time, he would have set up roughly 44 aquariums per month

---

45   Quoted in Bellanca, M.E., "Recollecting Nature: George Eliot's 'Ilfracombe Journal' and Victorian Women's Natural History Writing", *Modern Language Studies*, 27 (1997) 3/4, pp. 19–36, pp. 22–3.

3 The domestic aquarium

with the great majority being established in the years 1855–60 (after which he started to encounter difficulties paying his suppliers). Of these, 3,500 failed and that failure was, of course, accompanied by the death of all the fish and marine animals that Lloyd had sold. And this does not include the many fish that died in transit either to Lloyd's shop or to his customers. But even Lloyd himself found it difficult at first and noted that his earliest attempts yield only a 10 per cent success rate which gradually rose to 30 per cent and then to 50 per cent. So even at a point when Lloyd felt that he was sufficiently experienced to maintain a healthy aquarium he still expected half his fish to die. As early as 1856, the author of "The Aquarium Mania" spoke of fish as "miserable victims to the aquarium mania" and concluded that:

> It is frightful to think what widespread destruction the aquarium has caused.

However, in early 1876, Lloyd performed the home aquarium craze one last service when he helped Gosse, who, perhaps ironically, had never had much success as an aquarist, set up what he described as a "show-tank" in his home at Marychurch. This was the home aquarium to end all home aquariums and was based on Lloyd's circulation principle. The tank was made of slate and half-inch-thick plate glass and held 50 gallons of sea water. It was set up in a lumber room in the servants' quarters, and directly above it was a slate cistern. Below, in a disused outhouse, Gosse had a reservoir dug out. This was lined with slate and had a wooden cover and, in June, it was filled with three cartloads of water from Oddicombe. A pump, made of steel, vulcanite and glass, had been commissioned by Lloyd from the specialist manufacturer Chedgey and arrived in September. Lloyd generously made this last tribute to the superiority of his system a gift. The pump was almost immediately connected and in October it was possible to start stocking the tank largely by dredging in Torbay. The system (in a moment reminiscent of Anna Thynne's use of her maid in early aeration experiments) was dependent on the muscle power of Gosse's gardener, James Chudley, who pumped for half an hour three evenings per week with a total of 675 strokes in each session. The success of this effort can be shown by the fact that in 1882 the original load of sea water was still crystal clear and capable of supporting healthy animals including a lobster, which Gosse greatly loved despite its habit of killing other inmates of the tank. Gosse recorded that the total enterprise (presumably not including the pump) cost £60, which as will be seen from the figures quoted above was at the upper end of the cost for an aquarium even taking into account that Gosse had the skills to construct his own tank and had free access, via his friend Arthur Hunt's boat the *Gannet* and his boatman, to dredges and, thus, to marine animals.

The 1895 *Post Office Directory* for London listed no aquarium dealers and in 1897 there were only five to serve the entire country.

4
She sells seashells

We know exactly what 5 October 1858 was like at Pegwell Bay in Kent. The sky was pale with a few dark clouds clearing away. The early morning sun was reddening the sky but the light was still bleaching out the colour of the land and sea. Donati's comet was visible as a pale but distinct streak of smudged light in the sky above the cliffs. The donkey-ride man was just setting up and presumably hoping for a few last customers before winter really set in. An artist was looking intently at the stratification of the rocks that towered over the beach, perhaps considering the challenge to faith set by the new sciences of geology and palaeontology. Three well-dressed women and a child were trying to enjoy their holiday. The child was looking for something to do; two of the women were looking intently at an object – a stone or shell – that one has found on the beach, while another carried a basket perhaps stocked with specimens they had already found. Four other people and a dog are scattered along the beach, two of them also possibly engaged in some sort of study of the rocks or the shells or the seaweeds.[1]

We know all this so precisely because the Pre-Raphaelite painter William Dyce set it down as an almost forensic record of the conditions

---

1   On this painting, see Holmes, J., *The Pre-Raphaelites and Science* (New Haven, CT: Yale University Press, 2018), pp. 30–6.

that prevailed that day in his painting *Pegwell Bay, Kent – a Recollection of October 5th, 1858.*

The seaside that Dyce depicts is what might be termed the scientific seaside and his vision contrasts markedly with that of William Powell Frith in his painting *Life at the Seaside* (also known as *Ramsgate Sands*). Here a great heap of Victorian life, seen from the sea, disports itself. There are minstrels, animals and all manner of entertainment both public and private. Frith's seaside is what the fictional character Molly Miggs described in her Norfolk dialect:

> It was more like a fair than a baach. There wor in a manner o' spaaking thousands upon thousands o' paple there o' all sorts and sizes.[2]

In Frith there are all kinds of dynamic interactions in the crowd, which contrasts strikingly with the austerity of Dyce's picture. Here, as Timothy Hilton has observed, the figures in the foreground:

> stand together as a family group, yet not a group, for they stand apart from each other and nobody looks at anyone else.[3]

By the 1850s the seaside was well established as a part of the social landscape and not merely a place for more or less intrepid travellers or invalids and convalescents (although even in 1861 the guide to seaside resorts and spas, *Where Shall We Go?*, contained "remarks on climate" which were "addressed chiefly to invalids").[4] And what we see in these two paintings is a contest for the meaning of this emergent space. For Dyce it is a space for contemplation and study. For Frith it is a space for enjoyment. The scientific seaside of isolated individuals is contrasted with the seaside of fun and community. In Frith, the sea is not actually visible except at the very edge (the literal and littoral seaside), whereas in Dyce a landscape, seascape and skyscape with figures offers a

---

2   Spilling, J., *Molly Miggs's Trip to the Seaside* (London: Jarrold & Sons, n.d., 1870s), p. 68.
3   Hilton, T., *The Pre-Raphaelites* (London: Thames & Hudson, 1970), p. 129.
4   Black, A. and C., *Where Shall We Go?* (Edinburgh: A. & C. Black, 1866), p. ix.

comprehensive vision of a created world waiting to be explored. A space that is more liminal than littoral. And while in Frith we look inland at human life, in Dyce we look outward and upward at the manifestations of nature, a viewpoint that may explain the feature noted by Hilton whereby the human gaze is directed only at nature and not at other people. Perhaps it should come as no surprise that the great Victorian poem of faith and doubt, Matthew Arnold's *Dover Beach*, which was published in 1857 so could have been running through the minds of the people in Dyce's painting, should have been inspired by the seashore.

In thinking about Victorians and sea creatures, we should consider not only aquariums, public and private, but also various other manifestations of interest in marine life and the marine sciences, especially as these were experienced and enacted within non-professional settings. Just as the home aquarium contrasts with the Fish House, so ladies like those in Dyce's painting contrast with the gentlemen of the universities and the scientific institutes and men like Gosse who made their living from the popularisation of the study of the creatures and artefacts to be found in abundance along the British coast and in the inshore waters. And while the Fish House and the aquarium offered novelty and sensation, the sight of seashells and seaweeds had been available for a long time and their study and their place in the world of things was already well matured when fish came on the scene. What changed was that these interests were now contextualised by a very different understanding of marine science achieved through the unique visibility of the workings of the underwater world afforded by the plate-glass and slate fish tank. This chapter explores those encounters with fish and marine science outside the aquarium in the domain of managed nature – the scientific seaside, the riverbank, the marine science institute and the fish farm.

Of course, the idea of the seaside and, specifically, the scientific seaside, was by no means an invention of the 1850s. In 1811, Mary Anning of Lyme Regis had created a sensation by her discovery of a gigantic fossil "crocodile" and through the next decades she became a celebrity in the new science of palaeontology that was, of course, a central battleground in the growing struggle between scientific geology and creationism. But aside from this high-level squabble, Lyme soon became a place where people wanted to go to look at stones, to chisel out fossils or to buy specimens from Mary Anning's fossil shop.[5] At this early stage, women like Anning's friends the Philpot sisters (Elizabeth, Mary and Margaret) were prominent and respected members of the nascent palaeontological community as were Elizabeth Gray, who pursued her studies and collection on the coast at Girvan in Scotland, and Eliza Gordon-Cumming, who specialised in fossils of fish. In Torquay, Mary Wyatt was running a shop selling seashells and other similar curios in the 1830s. Hers was not the only shop of this kind operating in a growing resort: on the quay was J. Heggerty's (the proprietor also acted as one of the official guides to the nearby fossil-rich Kent Cavern). Heggerty's shop was known as an "extensive establishment", selling minerals and fossils. Wyatt also sold these (including polished specimens) as well as "specimens of the marine botany and shells of the Torbay district". Between 1834 and 1840, Mary Wyatt published the multi-volume *Algae Danmoniensis*, an album that contained 50 samples of actual seaweeds, gathered on the Devon coast, and preserved by pressing dry.[6] This she prepared under the guidance of Mrs Amelia Griffiths (in whose house she had once worked as a servant), a very serious phycologist who was praised as "worth ten thousand other collectors" in W.H. Harvey's textbook *Manual of the British Algae* (1841) and who had the seaweed *Griffithsia corallina* named after her by the great Swedish scientist Jacob Georg Agardh. Unfortunately, Mrs Griffiths published little herself so is remembered more as an influential

---

5   On Mary Anning, see Pierce, P., *Jurassic Mary* (Stroud: The History Press, 2006) and Emling, S., *The Fossil Hunter* (New York: St Martin Griffin, 2009). Given the foundational importance of Mary Anning to palaeontology, it is surprising that she has not been the subject of a full academic monograph.

6   Wyatt, M., *Algae Damnonienses, or Dried Specimens of Marine Plants, Principally Collected in Devonshire; Carefully Named According to Dr Hooker's British Flora. Prepared and sold by Mary Wyatt, dealer in shells, Torquay*, 5 volumes (Torquay: Cockrem for the author 1834–1840). Amelia Griffiths contributed the chapter on the "Natural History of Torquay, Including Marine Botany and Conchology" to Octavian Blewitt's *Panorama of Torquay*, 2nd edn., (London: Simpkin & Marshall and Torquay: Cockrem, 1833) which also provides some detail on Heggerty's and Wyatt's shops.

collector and correspondent than as a scientist in her own right but, to show the importance of her contribution to marine science, she identified 234 species of seaweed in Devon.

Anna Atkins' *Photographs of British Algae* appeared in 1843 and the hauntingly beautiful blue of its cyanotypes of the fractal-like forms of seaweed was the first photographically illustrated book.[7] A few years later in 1859, William Grosart Johnstone and Alexander Croall's *The Nature-printed British Sea-weeds* offered another gorgeously produced set of images printed by William Bradbury from lead plates on which the impress of each specimen had been made with a rolling mill. This process produced a slightly embossed image which, when coloured, looked remarkably like an actual dried specimen.[8] Louise Lane Clarke's *The Common Seaweeds of the British Coast and Channel Islands* (1865) offered its purchasers genuine specimens of the plants described and depicted, so was a shortcut to building an album of your own.[9] So although anyone could go to a beach and pick up seaweed, the market could bear extremely expensive reference books for the dedicated and wealthy collector. In this kind of collection one owned both the original specimen and an image of it. The specimen would have cost you nothing except the trouble of collecting and processing it; the image would have cost you the equivalent of a week's wages for the woman who made your bed.

---

7   Atkins, A., *Photographs of British Algae: Cyanotype Impressions* (privately printed for the author, 1843). A modern edition is Schaaf, L.J. (ed.), *Sun Gardens, Cyanotypes by Anna Atkins* (New York: New York Public Library, 2018). See also Garascia, A., "'Impressions of Plants Themselves': Materializing Eco-archival Practices with Anna Atkins's *Photographs of British Algae*", *Victorian Literature and Culture*, 47 (2019), pp. 267–303. On the popular hobby of seaweed pressing, see Haddad, A., "Nature for Ladies: The Victorian Art of Flower and Seaweed Pressing", Merchant's House Museum (https://merchantshouse.org/blog/seaweed-pressing, 21 August 2017).
8   Croall, A. and Johnstone, W.G., *The Nature-printed British Sea-weeds*, 4 vols (London: Bradbury & Evans, 1859–1860).
9   Clarke, L.L., *The Common Seaweeds of the British Coasts and Channel Islands* (London: Frederick Warne & Co., n.d. 1865?).

David Allen argues, I think correctly, that "the real growth of enthusiasm for marine life seems to date from the 1820s".[10] Certainly, by 1835, Mary Roberts' *Sea-side Companion* contained most of the ingredients to help an amateur marine scientist to enjoy a day on the beach.[11] In 1849 W.H. Harvey's influential *The Sea-side Book* appeared and the following year Anne Pratt's *Chapters on the Common Things of the Sea-side* and although Gosse and Kingsley are now seen as the most influential writers in the popularising of marine science that is, I think, the effect of their reputation in hindsight and not necessarily an accurate picture of who was in the pockets and hampers of amateur scientists on the Victorian beach.[12] So, the explosion of interest in the 1850s and beyond was rooted in an existing field of study that had already started to grow and, in fact, at the decorative level at least, seaweed had been used in silk and calico dress fabrics since the 1780s. For example, the designer William Kilburn devised at least 20 fabrics with seaweed motifs.[13] And in the Brighton Pavilion:

> Le Régent, future George IV, fit executer un ensemble des chaises et de tables dont la forme était celle de coquillages, les assises des chaises consistant d'une grande coquille sculptée dans le bois et peinte de façon à imiter la forme naturelle d'une coquille Saint-Jacques.
>
> [The Regent, the future George IV, had made a suite of tables and chairs of which the form was that of shells, the seats of the chairs consisted of a large shell sculpted in wood and painted so as to imitate the natural form of a great scallop.][14]

---

10 Allen, D., "Tastes and Crazes", in N. Jardine, J.A. Secord and E.C. Spary, *Cultures of Natural History* (Cambridge: Cambridge University Press, 1996), pp. 394–407, p. 397.
11 Roberts, M., *The Sea-side Companion* (London: Whittaker & Co., 1835).
12 Harvey, W.H., *The Sea-side Book* (London: John van Voorst, 1849); Pratt, A., *Chapters on the Common Things of the Sea-side* (London: SPCK, 1850).
13 See Christie, A., "A Taste for Seaweed: William Kilburn's Late Eighteenth-century Designs for Printed Cotton", *Journal of Design History*, 24 (2011), pp. 299–314.
14 Mauriès, P., *Coquillages et Rocailles* (Paris: Thames & Hudson, 1994), p. 75.

4 She sells seashells

The shell and seaweed craze (and the general marine science craze) made itself particularly visible in the design of Victorian majolica ceramics. High-end manufacturers such as Minton and Wedgewood (especially with its Ocean Pattern service) encrusted their wares with shells and crustacea. And majolica platters crawling with sea life became a common production of many less famous potteries from the late 1860s through to the end of the period and sat on the table or sideboard like little ceramic rockpools.

The interest in littoral sciences appears to have been associated with women from an early period. In 1808 the publisher of Dawson Turner's *Fuci* asked if the first part could be hurried to be released in April "so that it might come out in time for the ladies at the seaside". Harriet Campbell Sheppard, a conchologist in Quebec, published her work on Canadian shells as early as 1829.[15] As phycology and conchology were already linked to women, we can see that the specific associations of the home aquarium with women are clearly related to an existing gendering of marine science. In addition to Wyatt, Griffiths and Atkins, there were the pioneering Irish collector Ellen Hutchins; Isabella Gifford, author of *The Marine Botanist* (1840); and also, in Ireland, Anne Ball (the sister of Robert Ball, founder of the Aquatic Vivarium in Dublin) who collaborated with William Harvey on his *Phycologia Britannica* and after whom he named the genus *Ballia* and the species *Cladophora balliana*. And, perhaps the most significant of all, Margaret Gatty, author of *British Sea-weeds* (1863).[16] She started her illustrious career in seaweed as an invalid sent to recover from the

---

15 Sheppard, H.C., "On the Recent Shells Which Characterise Quebec and Its Environs", *Transactions. Literary and Historical Society of Quebec*, 1 (1829), pp. 188–97. See Creese, M.R.S. and Creese, T.M., *Ladies in the Laboratory III: South African, Australian, New Zealand and Canadian Women in Science: Nineteenth and Early Twentieth Centuries* (Lanham: Scarecrow Press, 2010).

16 See Hunt, S.E., "Free, Bold, Joyous: The Love of Seaweed in Margaret Gatty and Other Mid-Victorian writers", *Environment and History*, 11 (2005), pp. 5–34 and Plaisier, H., Bryant, J.A., Irvine, L.M., Jones, M. and Spencer Jones, M.E., "The Life and Work of Margaret Gatty (1809–1873), with Particular Reference to Her Seaweed Collections", *Archives of Natural History*, 43 (2016) pp. 336–50. Gatty's *British Sea-weeds* included detailed instructions for preparing and organising specimens.

birth of her seventh child by taking the sea air at Hastings. Thomasina Campbell, an eccentric expatriate resident of Corsica and friend of Edward Lear, passed her time by studying the island's flora and fish, and published her findings together with much else in her *Notes on the Island of Corsica in 1868*, which was reviewed warmly but eccentrically in *The Spectator* (28 November 1868):

> She describes the fish as "especially good." We are not quite sure we can implicitly trust Miss Campbell in this matter; we find her saying on one occasion that the diligence stops at a certain place for breakfast, but that "breakfast will seem to a botanist a waste of time."

There was also an anonymous lady in Tamworth who read her paper on "Seaweed" to the local Natural History Society (*Tamworth Herald*, 27 April 1871).

We have seen how Harvey praised Amelia Griffiths. A similar appreciation of the work of women in the field of phycology can be found in the introduction to Richard Greville's *Algae Britannicae* (1830), which was quoted with approval by *The Lady's Newspaper* (6 March 1853):

> It is not without a feeling of extreme pleasure ... that by means of this present work, I shall place in the hands of my fair and intelligent countrywomen a guide to some of the wonders of the great deep; nor need I be ashamed to confess that I have kept them in my view during the whole undertaking. To them we are indebted for much we know on the subject. The very beauty and delicacy of these objects have often attracted their attention.

It is noteworthy that Greville's acknowledgement of the quality of the female collectors whose scientific acumen had helped establish the field in which he now worked is tempered somewhat by his sense that there was also an aesthetic element in the world of phycology which might be more interesting to women than to men. And this perception is not helped by comments in a review of George Johnston's *An Introduction* to *Conchology* published in *The Lady's Newspaper* about five years later (13 July 1858):

> Such a book on such an enticing subject, has not till this time appeared. We care not whether the reader who takes it up likes shells or not; let her, for we presume we only address ladies, only begin and we venture that she (or he, should she call her husband, son, brother, or sweetheart) will not stop reading till she gets to Letter XXVII, when there are only eighty pages to go.

So, as well as drawing high praise from individual male scientists, the seriousness with which women approached the study of algae and shells can also be constantly undercut or framed by levity and an assumption that the attraction of the field is as much about the lure of prettiness as the rigour of taxonomy.

But it is clear that from the 1850s onwards there was a considerable boom in interest in the marine sciences, especially conchology and phycology, of a quite different order to the prehistory sketched above and that this complemented the aquarium craze that was itself, as we have seen, closely associated with women. It should be said that while women like Margaret Gatty and Amelia Griffiths were hugely admired by their male counterparts, they did not have access to the same institutional positions or facilities that the men had and perforce existed as a caste of gifted amateurs which made major contributions to the understanding of seaweed in particular. I think that the specific interest of women in shells and seaweeds was related to this exclusion as the raw material was to hand and did not require exotic expeditions or lengthy field trips and the equipment needed to analyse and evaluate the specimens would have been well within the reach of a well-off single woman or a married woman whose husband was prepared to support her interests.

However, there were other reasons for this gendering. One was the study of shells and seaweed did not involve the focus on sex and reproduction that botany required. Even so, as Kate Teltscher points out in her excellent work on the early history of Kew Gardens, Victorian views of the gardens, which were a site of science as much as recreation, almost always show many more female visitors than male.[17] And Helen

Paterson's engraving "Gathering Ferns", which appeared in the *Illustrated London News* (1 July 1871) at the height of pteridomania, shows eight women and three men. Two of the men appear to be there merely to carry things but the third stands at the centre of the picture inspecting a specimen that one of the ladies has brought to him, perhaps as a representation of deference to supposedly superior male knowledge of the field. And, of course, there were exceptions to the idea that botany was not quite right for women such as Jane Loudon's *Botany for Ladies* (1842), Elizabeth Twining's *Illustrations of the Natural Orders of Plants* (1849–55), and Sophy Moody's beautifully illustrated *The Palm Tree* (1864), which contains what I believe to be the archetypal image of the desert island – a patch of sand surrounded by the sea and sustaining a single palm tree.[18] There were also aspects of phycology and conchology that seemed specifically suited to women when seen as the fairer and weaker sex. As *The Naturalist* (19, 1864) put it when speaking of conchology:

> [It is] a study particularly suited to ladies; there is no cruelty in the pursuit, the subjects are so brightly clean, so suited to the boudoir.

The *Morning Post* (25 February 1843) made a very similar point but rather less kindly in its review of Agnes Catlow's book *Popular Conchology*:

> A little coquetry, however, with them, as with everything else is allowable, and when kept within feminine bounds, their gentle flirtations with geology, theology, astrology, and all the other ologies, are amusing to watch, and wholesome to laugh at. Conchology is one of the most harmless branches of natural history that a lady may dabble with.[19]

---

17 Teltscher, K., *Palace of Palms: Tropical Dreams and the Making of Kew* (London: Picador, 2020).
18 See Shteir, A.B., "Gender and 'Modern Botany' in Victorian England", *Osiris*, 12 (1997), pp. 29–38 (and various other works by this author); Dias, R. and Smith, K., *British Women and Cultural Practices of Empire 1770–1940* (London: Bloomsbury, 2018); and King, A.M., *Bloom: The Botanical Vernacular in the English Novel, 1770–1900* (Oxford: Oxford University Press, 2003).

4 She sells seashells

We may compare that sentiment with the more approving tone of the *Morning Chronicle* (9 October 1852) reviewing Mary Roberts' *Popular History of Mollusca*:

> Shells it appears to the writer, are too often treated as merely objects of ornament and fancy.

The review recognises the author's claims to have her scientific work taken seriously.[20] This is just under ten years after the appalling sentiments of the *Morning Post* review cited above and I think conveys not only a general shift in attitudes but a specific recognition of the work of many women marine scientists that could not be written off or ignored or wholesomely laughed at. In fact, there was a good deal of cruelty involved in conchology. It is true that shells collected from the beach were only the residue of lives that had once been lived and there was no harm in taking them. But professional shell collectors and shell dealers did take shells with the molluscs still in them and either prised them out or killed them by opening them or boiled them shell and all so that they could be easily removed. Hanneke Grootenboer, commenting on the miniature shells in the Wünderkammer of a seventeenth-century Dutch doll's house, points out that:

> The tiny shells are remnants of deceased baby sea creatures whose deaths were premature. They offer evidence of lives cut short, future potential denied.[21]

---

19 Catlow, A., *Popular Conchology, or the Shell Cabinet Arranged According to the Modern System* (London: Longman, Brown, Green & Longmans, 1854). This work included detailed instructions for the organisation of shell collections.
20 In fact, this work (*A Popular History of the Mollusca*, London: Reeve & Benham, 1851) was explicitly connected to Mary Roberts' much earlier *The Conchologist's Companion* (London: G. & W.B. Whittaker, 1824) and sought to link the two sciences of conchology and malacology and show their relationships.
21 Grootenboer, H., "Thinking with Shells in Petronella Oortman's Dollhouse", in M.A. Bass, A. Goldgar, H. Grootenboer and C. Swan, *Conchophilia* (Princeton, NJ: Princeton University Press, 2021), pp. 103–26, p. 104.

Modern shell dealers use microwave ovens to pop the animals out.

One of the most popular Victorian shell books, George Sowerby's *A Conchological Manual*, which first appeared in 1839 and went through four editions by 1852, dealt with the problem head on in the introduction to the 1852 edition. Sowerby largely addresses a male enthusiast (although he does refer to "the cabinet and the boudoir" as the sites of domestic conchology) and admits that the most scientifically rigorous approaches to shells are those that include the study of the molluscs that inhabit them. However, he is also pragmatic, adding that:

> At the same time, it must be admitted that there are many private collectors of Shells who would find it a difficult, if not impossible task, to study minutely and successfully, the soft parts of the Mollusca. Ladies, for instance, could not be expected to handle with pleasure and perseverance, these bodies, which in order to be preserved from putrefaction, must be kept in spirits; and yet such persons may, with improvement and advantage to their own minds, enjoy the interesting and scientific amusements of studying and arranging the clean and beautiful shells, which are so easily preserved, and so exquisitely beautiful in their structure.[22]

Some 20 years later a review of James Harting's *Rambles in Search of Shells, Land and Freshwater* in *Bell's Life* (4 September 1875) shows that perceptions had shifted even further:

> There was a time when the study of conchology was pursued with an immense amount of assiduity by private individuals of both sexes, and it was quite a common topic of conversation principally among such persons whose health forbade them to indulge in exercise of a severe nature, or those participated in by friends of a robust temperament. Ladies used to delight in explaining the diversity of their specimens, and took a pleasant pride in exhibiting

---

22 Sowerby, G., *A Conchological Manual* (London: H. Bohn, 1852), reprinted as *Sowerby's Book of Shells* (New York: Crescent Books, 1990), p. 12.

the various examples in their possession. A chief ornament of the morning room or boudoir was some fine specimens of either univalves or bivalves, as the object might prove to be ...There is a charm in the study of shells which, as the author of this small volume very justly remarks, commends itself to ladies who, whilst cultivating botanical researches, might at the same time associate with them a distinctive acquaintance with the tribe of British land mollusca.

The review also points out that the growth of science as a professional institution has gradually driven out the amateurs, competent and gifted though they may very well have been, and this in itself would have led to a reduction in the public recognition of female marine scientists who were doomed to remain outside the universities and the laboratories of learned societies.

There were exceptions of course: the brave and independent spirits who had the means to pursue their interests without reference to the social frameworks of gender, as the following notice from *Hearth and Home* (7 July 1898) shows:

Miss Hastie is about to undertake a scientific expedition to the South Seas ... Miss Prince, a well-known botanist will accompany Miss Hastie. Conchology and anthropology will also be studied.

This was a well-resourced expedition funded by Jane Alexia Hastie who had chartered the schooner *Sydney Belle* and secured the services of a group of male scientists as well as the Bostonian Frances Prince who is mentioned in the notice. Miss Hastie was a "Scottish lady of means" and had been in New Zealand for a time before she commissioned this expedition (*Wanganui Herald*, 21 March 1898). Problems with the ship led to the project being abandoned although Miss Hastie successfully sued the owners of the *Sydney Belle* for breach of contract (*Wanganui Chronicle*, 24 February 1899).[23] By 1900 she was fly-fishing in the far

---

23 I am indebted to the New Zealand citizen scientist Siobhan Leachman for getting me back on the track of the redoubtable Miss Hastie when I thought the trail had gone cold.

north and was pictured in action in a wonderful photograph in William Bisiker's *Across Iceland*.[24] Another well-travelled independent woman at much the same time was the botanical illustrator (including some images of fish) and plant hunter Marianne North, who went everywhere and was photographed by Julia Margaret Cameron wearing local dress at her house in Ceylon. Notice that both Hastie and North were single women of considerable private means (Marianne North was sufficiently wealthy to give Kew Gardens her paintings and the money to build a gallery to keep them in). Sarah Bowdich Lee was doing fieldwork in West Africa with her husband as early as 1816 and, after he died there, she went on to make a name for herself as an ichthyologist in her own right.[25] And there was also Mary Kingsley, author of *Travels in West Africa* (1897), who brought home from her two expeditions 65 different species of fish, two of which were named for her. As she remarked of herself:

> I can honestly and truly say that there are only two things I am proud of – one is that Doctor Günther [author of the *Introduction to the Study of Fishes* (1880)] has approved of my fishes and the other is that I can paddle an Ogowé canoe.[26]

Women also participated in large numbers in the public science of the day. For example, *The Belfast News-Letter* (21 January 1852), reporting on a public lecture on conchology by Richard Davidson, commented positively that:

---

24   Bisiker, W., *Across Iceland* (London: Edward Arnold, 1902). The photograph of Miss Hastie may be found on p. 130.
25   See Orr, M., "Women Peers in the Scientific Realm: Sarah Bowdich (Lee)'s Expert Collaborations with Georges Cuvier, 1825–1833", *Notes and Records of the Royal Society*, 69 (2015), pp. 37–51; Gates, B., *Kindred Nature* (Chicago: Chicago University Press, 1998); Orr, M., "Fish with a Different Angle: *The Fresh-Water Fishes of Great Britain* by Mrs Sarah Bowdich (1791–1856)", *Annals of Science*, 71 (2014), pp. 206–40.
26   Kingsley, M., *Travels in West Africa* (London: Macmillan, 1897), p. 63.

4 She sells seashells

> Notwithstanding the inclement state of the weather, the audience was very numerous, the room being quite filled, and a large proportion being ladies.

One challenge for serious lady scientists and collectors of both seaweed and seashells was the constant pressure to treat the objects of their interest as terms in a grammar of domestic aesthetics. At the lower end of the scale, objects made of seashells or decorated with seashells were a common feature of Victorian clutter and making them was an approved pastime for ladies. Similarly, pictures made of dried seaweed competed with the more serious seaweed albums of the amateur collectors for the attention of the public. Here, a fashionable journal, *Le Follet: Journal du Grand Monde, Fashion, Polite Literature, Beaux Arts &c. &c.* (1 January 1875), advises its female readers on how to combine science and domesticity:

> Seaweed and shells, of every variety of pattern or design, are left ashore by each receding tide, so that it is impossible to take a walk along the beach without finding something new and beautiful ... [of shells] the collection that you make – those beautiful forms of the sea with which you store your home, may be handed down by you uninjured to your children.

Against this the Reverend James Gilbert's article "Lessons of the Shore", which appeared in *The Girl's Own Paper* in 1880 or 1881, linked the shapes of shells with the more serious business of understanding that shape in the context of the understanding of the good order of divine creation, which was frequently the aim of Victorian science:

> When, therefore, we pick up a shell, we should instinctively ask ourselves, how the shell came to be of that particular form, and why that form should have been chosen.

The religious distinctions also worked alongside the gendering of marine sciences. As Fabienne Moine points out:

Victorian Quakers, including a significant number of women, were thus attracted to the empirical branches of observational science such as botany, ornithology, horticulture, entomology, and conchology, all innocent and purposeful occupations compared with the manufacture of weapons or inappropriate pursuits such as music.[27]

The seaweed craze lasted a long time. In 1885 the magazine *Decorator and Furnisher* (6) could still run an article on "Decorative Work for Ladies" that included material on drying and arranging seaweed while, as late as 1895, a piece by H.L. Jelliffe entitled "The Artistic Value of Seaweed" appeared in the *The Monthly Illustrator* (5) and demonstrated that the connection between the aesthetic pleasure of seaweed and its scientific study had still not been severed:

> Who can withstand the temptation to put some of these filmy things in little vials of salt water, and carry them carefully home to be examined under the microscope, named and mounted as a perpetual reminder of the summer and the sea-gardens.

This last point perhaps offers another perspective on the crazes for both seashells and seaweed. Both were portable reminders of happy days, quite literally souvenirs, and while aquariums enabled the sea to be memorialised by being set up in model form in the parlour, shells and algae represented tangible evidence of your presence at a particular time and place. You selected them with your eyes and then bent to pick up the chosen specimen that you would then take home or back to your hotel or guesthouse to prepare. To facilitate this, shops at seaside resorts often sold albums for mounting specimens of seaweed, packets of Bentall's Botanical Drying Paper and preprinted botanical labels. As late as 1893, handbooks such as Hervey's *Sea Mosses* contained detailed instructions for the drying and mounting of specimens. So, the collections of shells and algae enabled the same pleasures as might be had by browsing a photograph album of holiday snaps. This is not

---

27   Moine, F., *Women Poets in the Victorian Era: Cultural Practices and Nature Poetry* (London: Routledge, 2016), p. 220.

a trivial function and speaks to profound human desires for the reconstruction and continuation of experience, especially pleasurable experience, as a bulwark against decay and mortality.

But these gentle and harmless pursuits could also be the occasion of violence. When Patrick Daniel's wife would not give him a penny for a shave, a row ensued which ended when she threw a seashell that hit him on the head and killed him (*Hereford Journal*, 2 May 1863). Apart from Mrs Daniel's marvellously strong arm and unerring aim, it is interesting that a seashell should be ready at hand in a household where money was apparently very tight and the sea was relatively distant.

Shells were also big business and, as Sowerby points out, with a peculiarly Victorian blend of capitalist practicality and idealism, without that business the frontiers of marine science would never have been pushed back:

> Let it also be remembered, that if shells had not been rendered commercially valuable, by the zeal and emulation manifested by *mere* Conchologists for the possession of rare specimens, few travelling merchants and sea captains would have thought them worthy of a corner in their cabins. In this case, few specimens being brought to the country, the more Philosophical Naturalist would have been without the means of obtaining materials to work upon, or of attracting public attention to his favourite pursuit.[28]

There were wholesale shell dealers in Liverpool and London and these businesses served not only collectors but also the fashion industry. One company in California shipped 40 tons of shells to France every two months at a price of $700–$1,000 per ton and these were made into buttons and shawl clasps and then shipped back to the USA. The wild animal dealer Charles Jamrach was an important trader in shells and in 1833 Marcus Samuel opened his first shell shop in Houndsditch. This prospered through the century and eventually prospecting for shells brought him into the oil business (which is why Shell Oil is so called

---

28 Sowerby, G., *A Conchological Manual* (London: H. Bohn, 1852), reprinted as *Sowerby's Book of Shells* (New York: Crescent Books, 1990), p. 12.

and why it has a shell as its logo). Although there was never the bubble in prices for shells as there had been during the Dutch and French shell manias of the seventeenth and eighteenth centuries, individual shells could still command very high prices. In 1825, for example, a specimen of *Conus gloriamaris* sold for £105 in the auction of the Tankerville collection and the auction of the Dennison collection in 1865 included many shells which sold for £40 or above. In 1846 the British Museum turned down the offer of Hugh Cuming – by far the most important Victorian conchologist and known as "the Prince of Collectors" – to take his collection for £6,000 but bought it (by now grown to 83,000 specimens), from his estate, for the same price in 1866. Cuming had a network of collectors that included the young Philip Gosse when he was in Jamaica. It should be noted that the shells had been collected by Cuming mainly while the animals that lived in them were still alive (hence he could be very accurate about identification), so this massive repository represented a great deal of ecological damage and animal cruelty.[29] Henry Mayhew noted that while Sowerby paid 30 guineas for a damaged specimen of *Conus gloriamaris* and an undamaged specimen would have cost 100 guineas, a network of about 50 peddlers of shells tramped routes reaching as far as Bristol and Liverpool and collectively bought more than one million common shells each year, paying 3s per gross and selling for 1d each. The attraction of seashells was such that a Sea Shell Mission, from 1888 under the patronage of Princess Mary of Teck, was founded to give poor urban children boxes of shells collected by "more fortunate boys and girls who visit or reside at the sea-side" (*Young England*, 1 September 1886).[30] Perhaps another member of the royal family may have contributed:

---

29 See Dance, S.P., *Shell Collecting* (London: Faber & Faber, 1966) for a comprehensive account of Victorian shell collectors. For earlier approaches, see Delbourgo, J., *Collecting the World* (London: Penguin, 2018). See also Dance, S.P., "Hugh Cuming (1791–1865), Prince of Collectors", *Journal of the Society for the Bibliography of Natural History*, 9 (1980), pp. 477–501.

30 The connection between the royal family and seashells – and an illustration of the ubiquity of shells in English life – can be seen in Lady Elizabeth Keith Heathcote's very pretty 1822 drawing of Princess Victoria playing with seashells in Ramsgate. Murphy, D., *The Young Victoria* (New Haven: Yale University Press, 2019), p. 67.

4 She sells seashells

A prince was on the beach at Osborne when he came across a boy with a basket of sea shells which he deliberately upset. The boy picked them up when the Prince upset them again. The Prince got such a licking as few Princes ever had!

(*Chums*, 31 March 1894)

The value of shells, even at a penny each, made them worth stealing and the *Hereford Mercury and Reformer* (13 February 1858) reported that Thomas Shaw and James Punter were in court for stealing a thermometer, a seashell and two puddings from Mrs Dean's house. Shaw was sentenced to a month in prison while Punter was discharged.

Seaweed had an economic value and various attempts were made to turn it into usable fibres. It was also a foodstuff on some coasts (in South Wales it is still easy to buy in Swansea market) and this led to severe unrest in part of Ireland during the various hungry years before and after the Great Famine. There was rioting over the right to collect seaweed to eat and some landowners took various villagers to court for trespass and damage – even though it appears they did not want the seaweed themselves and only left it to rot. In one case reported by the *Cork Examiner* (21 April 1843), the trial was suspended when the bench ruled that the landowners had no good reason to prevent seaweed collection and that, in any case, cutting seaweed below the low water mark or collecting it from a rowing boat inshore could not be deemed trespass. As late as 1883, a "Seaweed Starvation Fund" had to be set up to alleviate famine in rural Ireland where many people had only seaweed to eat. The genre of Victorian paintings showing seaweed collectors are usually sanitised images of happy peasantry and fisherfolk harvesting an unusual crop but, in this case as in others, we need to look at what John Barrell memorably referred to as "the dark side of the landscape" and the poverty and misery that could underpin this activity.[31]

---

31 Barrell, J., *The Dark Side of the Landscape* (Cambridge: Cambridge University Press, rev. edn. 1983). See also O'Connor, K., *Seaweed, A Global History* (London: Reaktion Books, 2017), pp. 116–20.

Another marine material, coral, also enjoyed a vogue in Victorian England. Coral had played a most important role in the early theories of Darwin, and Anna Thynne reported her observation of coral reproduction. In addition, Joseph Beete Jukes, in his *Narrative of the Surveying Voyage of HMS Fly* (1847), commented on the amazing structures of coral reefs.[32] The whole notion of tiny coral organisms building mighty structures caught the Victorian imagination, whether in the poetry of Richard Garnett or James Gates Percival, musical pieces such as *The Coral Waltzes*, commentary by Dickens or the use of this setting for Ballantyne's didactic tale *The Coral Island* (1857), which included commentary on the place of coral in the divine scheme and the wonders of God-given coral architecture. As Katherine Anderson has pointed out:

> Corals were ambiguous organisms, straddling the borders of plant and animal life, as their classification, zoophytes (*zoo*, "animal"; *phyte*, "plant") suggests. This ambiguity dated to the classical era, but it gained new meaning with early nineteenth-century studies of marine invertebrates (including corals, anemones, sponges, molluscs, and jellyfish).[33]

Coral became highly fashionable for a time as a component in luxury jewellery, at first elaborately carved and often imported from Naples and Genoa (but also produced in London – notably by the high-end Phillips Brothers' jewellery studio) and subsequently in more naturalistic form with branches of coral more simply set in precious metals. So even in the most expensive commodities, there was a memory of marine science and the brooches, earrings and necklaces of society ladies or the tie pins of gentlemen take their place in the transition of the sea and its creatures into the things that conveyed

---

32 See Shick, J.M., *Where Corals Lie* (London: Reaktion Books, 2018) and Davidson, K., "Speculative Viewing: Victorians' Encounters with Coral", in G. Moore and M.J. Smith (eds), *Victorian Environments: Acclimatizing to Change in British Domestic and Colonial Culture* (London: Palgrave Macmillan, 2018), pp. 135–60.
33 Anderson, K., "Coral Jewellery", *Victorian Review*, 34 (2008), pp. 47–52, p. 49.

4 She sells seashells

social meaning and that shift in the quality of looking that begins in the Fish House. The new vision of the sea and its creatures relocated these high-end items in the grammar of exchange and gave them new meaning as commodities. And, as Anderson says, the ability of the consumers of this material to make that new conceptual connection and the semantic shift that flows from it were determined by the work of Gosse and others and the experience of rockpools:

> If red coral jewellery was a Mediterranean product, then Victorian consumers had repeatedly been taught the significance of their modest counterparts on British shores.[34]

Developments in microscopy meant that water itself became the site of new ways of seeing. In 1851 the conchologist Agnes Catlow published *Drops of Water*, which showed what was actually to be seen in water when viewed microscopically.[35] This was fascinating but not very comforting when you had to drink it, as William Heath's cartoon "Monster Soup, Commonly Called Thames Water" (c. 1828) suggested: it shows a lady dropping her teacup in horror after she has peered through a microscope and seen the rich fauna swimming in a drop of the water she has just drunk. One of Mayhew's informants was a man who set up a microscope on the streets of London and charged people to look at various slides he had prepared, including one of a drop of water. And the microscope, as has already been suggested elsewhere, became a crucial element in the equipment of anyone who wanted to take their seaside holiday on the scientific rather than recreational beach. As the philosopher Herbert Spencer put it:

> Whoever at the sea-side has not a microscope and aquarium, has yet to learn what the highest pleasures of the sea-side are.[36]

---

34 Anderson, K., "Coral Jewellery", *Victorian Review*, 34 (2008), pp. 47–52, p. 49.
35 Catlow, A., *Drops of Water* (London: Reeve & Benham, 1851).
36 Spencer, H., *Education: Intellectual, Moral and Physical* (London: Williams & Norgate, 1861), p. 45.

The holiday makers depicted at Ramsgate by Frith don't look as if they would agree. Or as J.E. Taylor – who in 1872 described a seaside visit as "now almost an annual occurrence for most people" – suggested:

> You can just as well expect to be an astronomer without a telescope, as to be a naturalist without a microscope.[37]

And then there was the challenge of what to wear on the beach. Victorian women's clothes were not entirely practical for clambering over rocks and exploring tidal pools. Some managed it. George Johnston, the author of, among many other things, *An Introduction to Conchology* (1850), was accompanied on his collecting expeditions by the doughty Mrs Johnston whom, he recalled:

> carried a larger muff than the present fashion would commend and many a heavy stone and well-filled bottle has therein been smuggled.[38]

However, a few years later G.H. Lewes was recommending a quite specific beach outfit in his *Sea-Side Studies at Ilfracombe, Tenby, the Scilly Isles, and Jersey* (1858):

> Our costume was but indifferently adapted to the drawing room and would have obtained small suffrage on the Boulevard des Italiens, the Prater or Pall-Mall. You shall judge. We are a lady [this was George Eliot who accompanied her future husband on these expeditions and was as enthralled by seaweed and marine life as any Victorian] and two men. The lady, except that she has taken the precaution of putting on the things "which won't spoil" has nothing out of the ordinary in her costume. We are thus arrayed: a wide-awake hat; an old coat, with manifold pockets in unexpected places, over which is slung a leathern case containing hammer, chisel, oyster-knife, and paper knife; trousers warranted

---

37 Taylor, J.E., *Half Hours at the Seaside* (London: David Bogue, 1880), p. 20.
38 George Johnston to Margaret Gatty, quoted in Allen, D.E. *The Naturalist in Britain* (London: Allen Lane, 1976), p. 154.

not to spoil; *over* the trousers are drawn huge worsted stockings, over which again are drawn huge leather boots. Mine are fisherman's boots and come a few inches over the knee ... In this costume we wooed the mermaids. We brought a crowbar, to turn over heavy stones, which could not otherwise be moved but which are worth moving, because it is under such that rarities will be hidden.[39]

So, while the men dress in the clothes of the working class and carry the tools of manual labour, we see that George Eliot, a representative on this occasion of the Victorian lady beachcomber, is dressed as usual but in less sumptuous fabrics and we can see this very outfit worn by the women on the beach in Dyce's painting of Pegwell Bay. It is notable that, while for the men a field trip in the pursuit of marine science involves a transgressive shift in appearance and a temporary adoption of a different class identity, for the woman the transgression does not extend to a change of clothes, and where there is transgression, it is in her involvement in the world of science and, especially, scientific expeditions. And when groups of men did go onto the beach for amateur science, other kinds of social barriers might also be breached. A visit to the shoreline by Edward Lear in company with some of the Pre-Raphaelite painters ended with Lear in the role of domestic servant and carrying John Everett Millais' collection of cuttlefish shells for him. The littoral space is also a liminal space and such spaces are always inhabited by transgressive perils.

But the class-transgressive sartorial options offered by the shoreline were amplified by the class-transgressive nature of the work carried out

---

39 Lewes, G.H., *Sea-side Studies at Ilfracombe, Tenby, the Scilly Isles & Jersey* (London: William Blackwood & Sons, 1858), pp. 17–18. See also Bellanca, M.E., "Recollecting Nature: George Eliot's 'Ilfracombe Journal' and Victorian Women's Natural History Writing", *Modern Language Studies*, 27 (1997), pp. 19–36; King, A.M., "George Eliot and Science", in G. Levine and N. Henry (eds), *The Cambridge Companion to George Eliot* (Cambridge: Cambridge University Press, 2019); and Feuerstein, A., "Falling in Love with Seaweeds; The Seaside Environments of George Eliot and G.H. Lewes", in L.W. Mazzeno and R.D. Morrison (eds), *Victorian Writers and the Environment* (London: Routledge, 2016).

there. The middle-class zoologists chipping away with their hammers and chisels offer an almost parodic image of the genuine work done by shoreline-dwelling workers in search of food. Here is John Harper's description of French peasants harvesting the rock-boring molluscs Pholadidae or piddocks:

> At the present day they are largely used as an item of food in France and Italy and on the coasts of the Mediterranean, where they abound. In the neighbourhood of Dieppe, Mr Stark tells us that bands of women and children, each armed with a pickaxe, make a formidable array against the unhappy Pholades, who tremble in their rock-citadels as the besiegers approach. By means of the sharp point of this implement they are able to detach considerable fragments of the rocks and a rich harvest of the molluscs ensues.[40]

The hard work of the French women and children is, in Lewes' book, transformed into the leisure of English ladies and gentlemen. While the French were happily foraging for food, Matilda Sophia Lovell's *The Edible Mollusks of Great Britain and Ireland* was designed not only to inform its readers about the physiology and habitat of this class of animals but also to tell them how to cook them. She noted that very few kinds of shellfish are eaten in the British Isles although the coastline abounded in edible specimens. Lovell tells us that:

> The Normand method of cooking the Pholas (*le dail commun*) is to dress them with herbs and breadcrumbs or pickle them with vinegar.[41]

---

40 Harper, J., *The Seaside and Aquarium* (Edinburgh: William P. Nimmo, 1868), p. 86.
41 Lovell, M.S., *The Edible Mollusks of Great Britain and Ireland with Recipes for Cooking Them* (London: Reeve & Co., 1867), p. 7. The present book does not address the question of fish as food (except where it mentions acclimatisation). Anyone interested in this topic might start with Picard, L., *Victorian London* (London: Weidenfeld & Nicolson, 2005), pp. 150–51 then Gray, A., *The Greedy Queen* (London: Profile Books, 2018) and Bloomfield, A.L., *Food and Cooking in Victorian England* (Westport: Praeger Publishers

Even the pickling jars in the hotel rooms of amateur marine scientists like Lewes and Eliot appear to mirror and parody the working-class relationship with Pholadidae.

A few years later still, Margaret Gatty, a far more serious marine scientist than Lewes, gave the following detailed description of what a lady should wear when collecting on the seashore:

> About this shore-hunting, however, as regards my own sex (so many of whom, I know, are interested in the pursuit), many difficulties are apt to arise; among the foremost of which must be mentioned the risk of cold and destruction of clothes. The best pair of single-soled kid Balmoral boots that ever were made will not stand salt water many days – indeed, would scarcely "come on" after being thoroughly wetted two or three times in succession – and the sea weed collector who has to pick her way to save her boots will never be a loving disciple as long as she lives! Any one, therefore, really intending to *work* in the matter, must lay aside for a time all thought of conventional appearances, and be content to support the weight of a pair of boy's shooting boots, which, furthermore, should be rendered as far water-proof as possible by receiving a thin coat of neat's-foot-oil, such as is used by fishermen – a process well understood in most lodging houses. It is true that sea-water does not usually give any one cold, but in sea-weed hunting, where there is so much standing and dawdling about, as well as walking, it is as well for the hardier hunters who have learned to walk boldly into a pool if they suspect there is anything worth having in the middle of it, they will oil their boots, for the simple reason that it is a mere waste of time to black and polish them; for polish as they will, a saline incrustation is sure to steal through at last. This advice cannot be enforced too strongly. It is both wasteful, uncomfortable, and dangerous to attempt sea-weed hunting in delicate boots. Wasteful, because a guinea pair will scarcely last a week. Uncomfortable, because to walk on some rocks in thin soles (the slate edges of those in

---

Inc., 2007). Collingham, L., *The Hungry Empire* (London: Vintage, 2018) looks at the place of fish in the overall imperial food economy.

Douglas Bay, for example) is so painful, that it very soon becomes impossible. Dangerous, because you must be wetted through by the first bit of moist sand you come to, and it is not everyone who would be justified in running the risk involved in this fact.

Next to boots comes the question of petticoats; and if anything could excuse a woman for imitating the costume of a man, it would be what she suffers as a sea-weed collector from those necessary draperies! But to make the best of a bad matter, let woolen be in the ascendant as much as possible; and let the petticoats never come below the ankle. A ladies' yachting costume has come into fashion of late, which is, perhaps, as near perfect for shore-work as anything that could be devised. It is a suit consisting of a full short skirt of blue flannel or serge (like very fine bathing gown material), with waistcoat and jacket to match. Cloaks and shawls which necessarily hamper the arms, besides having long ends and corners which cannot fail to get soaked, are, of course, very inconvenient, and should be as much avoided as possible; but where this cannot be, a good deal can be done by tucking them neatly up out of the way. In conclusion, a hat is preferable to a bonnet, merino stockings to cotton ones, and a strong pair of gloves is indispensable. All millinery work – silks, satins, lace, bracelets, and other jewellery, etc., must, and will be, laid aside by every rational being who attempts to shore hunt.

A stick was alluded to before, and is a very desirable appendage, both as a balance in rock-clambering and for drawing sea-weeds from the water. It should have a crook for a handle therefore. But about these sorts of matters, people should amuse themselves by devising ingenious varieties. The basket may be lined with gutta percha, or exchanged for those who care to invest in it, for an Indian-rubber bag which can be strapped round the waist, and into an inside pocket of which a bottle or two for the more delicate sea-weeds may be easily stowed away. But the common basket which has served the previous generation will do very well for any one who is in earnest in this. Few tools come amiss to a good workman, and it argues a rather *dilettante* state of mind to insist on having everything the perfection of convenience. Into which question comes also that of expenditure;

and the reader is here assured, once for all, that it is quite possible to go shore-hunting without any extra expense whatsoever; that *very* strong-soled pair of boots excepted, and they will be found quite as useful in country walks afterwards, as on the sands.[42]

I don't think Mrs Gatty needed a male escort, especially not one as irritating as Lewes. But what we see here is an interesting progression from Mrs Johnston with her muff, through George Eliot, one assumes still bonneted and shawled, to Margaret Gatty kitted out for some serious marine science. She clearly knows that the easiest thing for her would be to wear men's clothes but that is a transgression too far so she adopts the most practical and professional approach she can and this was what won her the admiration of the institutionalised male scientists of the day. We can see an outfit very similar to the one recommended by Margaret Gatty in a sketch of Mary Anning at work, jotted down by her friend, the eminent geologist Sir Henry De la Beche, sometime in the late 1830s.[43] She has stout boots and what appear to be trousers peeping out from beneath the hem of her good thick skirt. She wears what looks like a top hat, tarred to make it weatherproof but also, I imagine, offering some protection against the falling rocks, which were always a hazard below the unstable cliffs of Lyme Regis – some of which killed her dog Tray when she was out with him on the beach. But Mary Anning was a working-class woman so her adoption of what are unambiguously working clothes is less transgressive than it would be for a lady like Margaret Gatty.

Gatty also corresponded with William Lloyd and purchased from his shop. Indeed, her first contact with him was to complain about the poor service she had received. Lloyd made amends (he had left his

---

42   Gatty, M., *British Sea-weeds. Drawn from Professor Harvey's "Phycologia Britannica"* (London: Bell & Daldy, 1872), pp. vii–viii. It's surprising that George Eliot didn't go further in modifying her dress for the beach but as Kathryn Hughes points out she was extremely fashion conscious (although she didn't always get it right). Hughes, K., *The Victorians Undone* (London: Fourth Estate, 2018) p. 184.
43   This is reproduced in the illustrations section between p. 110 and p. 111 in Pierce, P., *Jurassic Mary* (Stroud: The History Press, 2006).

brother to mind the shop) and, from then on, he and Mrs Gatty kept up a cordial and well-informed exchange.

The naturalist and successful aquarist Helen Watney, writing in 1871, describes a somewhat genteel approach to collecting:

> It was great fun to drive down [to Southsea] from my little den at Hambledon and spend the entire day on the shore collecting, and returning home in the cool of the evening, through the green Hampshire lanes, the pony carriage laden with jars of salt-water and the hamper full of sea-weeds – *the* hamper, which had in the morning been the receptacle for our sandwiches and sherry.[44]

If you didn't want to do the work yourself, there were local people who could do it for you. Gosse himself used such professional dredging services. In Weymouth alone you could employ the unfortunately named fisherman Jonah Fowler, William Thomson or Robert Dawson to go out in a boat and dredge the seabed for specimens for your collection, and other seaside towns had similar characters who were only too pleased to supplement their income by dredging for a gentleman or lady scientist.

The encounter between the professional male scientists and the amateur lady can be seen several times in the life of Philip Gosse. He refers to "Oddicombe, whither ladies repair to search for pebbles containing fossil madrepores" or to "a boat, painted in white and green, for the attraction of young ladies of maritime aspirations" and a vignette in the earliest biography shows him actually encountering these earnest women on whom his fortune (via the sale of his many books) was based:

> We came across a party of ladies, who were cackling so joyously over a rarity they had secured that our curiousity [sic] overcame our shyness, and we asked them what they had found. They named a very rare species, and held it up for us to examine. My father [Philip Gosse], at once, civilly set them right; it was a so-and-so, something much more commonplace. The ladies drew

---

44 Watney, H., "Marine Aquaria", *Hardwicke's Science Gossip*, 7 (1871), p. 196.

themselves up with dignity, and sarcastically remarked that it was the rarity, and that "Gosse is our authority."[45]

Just as the aquarium mania led to the cruel neglect and destruction of thousands of fish, so the British seashores were wrecked by the onslaught of the amateur naturalists and collectors. As early as 1832, James Clealand was reporting of the beaches around Bangor in Northern Ireland that:

> My Patellas are almost extirpated, they became so much the fashion that the visitors who frequented Bangor as Sea Bathers, during the last two summers, employed the children to collect them and there is not one to be seen now.[46]

Gosse recognised that the impact of his popular work *Tenby, a Seaside Holiday* had resulted in an ecological catastrophe for the very things of beauty he had so praised. As he wrote to William Lloyd in 1856:

> The caverns here are not what they were in 1854. Some cause or other, either the rapacity of amateurs, or the frosts of the winter 1854-5, has almost quite extirpated the Actiniae that were so abundant then ... I have not been able to hear of anyone who would undertake to collect for you as a business; and indeed thus

---

45 "Oddicombe", Gosse, P.H., *A Naturalist's Rambles on the Devonshire Coast* (London: John van Voorst, 1853), p. 21; "a boat", Gosse, P.H., *A Year at the Shore* (London: Alexander Strahan, 1865), p. 64; "a party of ladies", Gosse, E.W., *The Naturalist of the Sea Shore: The Life of Philip Henry Gosse* (London: Heinemann, 1890), p. 288. Oddly enough the young Huxley was dredging at Tenby in 1855 so just missed Gosse's famous holiday there. Huxley was on honeymoon and spent the days dredging and the evenings dissecting and classifying his catch while his new wife helped him write the descriptions (Desmond, A., *Huxley, The Devil's Disciple* (London: Perseus Books, 1997), pp. 211–15).
46 On Clealand, see MacDonald, R. and McMillan, N., "James Dowsett Rose Clealand (Cleland): A Forgotten Irish Naturalist", *Irish Naturalist's Journal*, 13 (1959), pp. 70–2. Clealand would have felt the loss of patellas most keenly as it was a distinctive species of patella discovered by him that gave him what small reputation he had.

denuding the caves and rocks of a town like this, of objects that afford such an inducement for visitors to come here, would be so unpopular, and indeed so unjust and selfish, that I would not on any account lend my aid to it. There are plenty of places in the vicinity, which are rich enough and quite unvisited; but I should be sorry to see Tenby and its caves robbed of their Actiniae.[47]

But the damage was already done and Gosse had the uncomfortable realisation that the destruction of the littoral environment by enthusiasts with one of his books in their hand and another in their pocket was partly his responsibility. In 1907 Gosse's son Edmund recalled his father's distress at what he had unwittingly caused:

> An army of "collectors" has passed over them, and ravaged every corner of them. The fairy paradise has been violated, the exquisite product of centuries of natural selection has been crushed under the paw of well-meaning, idle-minded curiousity [sic]. That my Father, himself so reverent, so conservative, had, by the popularity of his books acquired the direct responsibility for a calamity that he never anticipated, became clear enough to himself before many years had passed, and cost him great chagrin. No one will see again on the shore of England what I saw in my childhood, the submarine vision of dark rocks, speckled and starred with an infinite variety of colour, and streamed over by flags of royal crimson and purple.[48]

Helen Watney's 1871 account of maintaining two large specimens of the sea anemone *Bunodes crassicornis* demonstrates almost all of the features of this kind of collecting, relieved only by the fact that Mrs Watney was a very serious naturalist, knew all about keeping marine life healthy and alive in the home aquarium and took genuine and informed pleasure in her scientific pursuits:

---

47 Quoted in Thwaite, A., *Glimpses of the Wonderful* (London: Faber & Faber), pp. 187–8.
48 Gosse, E.W., *Father and Son* (London: Heinemann, 1907), p. 150.

4 She sells seashells

One was brought to me by a fisher lad; the other was found by my boy and a young friend of his, close to the oyster beds at the point [at Beaumaris], attached to a large piece of stone. The boys, very wisely, did not attempt to remove it from its mooring, but carried it between them, the miniature rock and all. It now stands in my sitting-room in the centre of a large brown pan full of sea-water which is constantly changed.[49]

Notice how a local boy has seen a lady collecting by the seashore and has done some speculative fieldwork of his own. I imagine that the rare anemone earned him a nice tip.

The author of "Books in the Running Brooks" (*Sharpe's London Magazine*, 17, 1860) could describe the shore of South Devon as untouched by the "hand of the spoiler in 1854" and observed that the shoreline rocks were:

> thickly encrusted with shelly molluscs of all kinds, and clothed with perfect gardens of rich-tinted brown, and green, and purple, and red sea-weeds, glorious to behold; whilst amidst the sparkling waters in the tide pools hosts of bright coloured anemones expanded their petal-like tentacles in safety, and delicate star-fish, and echini, and thousands of other animals glided about amongst the floating fronds of the tangles and ulvas which fringed their sides.

But a return visit in 1858 revealed a shocking change:

> I did not find *one single specimen of any description*, except a few of the more common mesembryanthemums. The stones for miles along the shore had been turned over and over, the rich crops of weeds and sponges that were so lovely utterly destroyed; the solid rock, wherever it has been disposed in shelves under which the waters could wash had been ripped up with crowbars and levers, and all the aged sires of the sea-cucumber, sea-anemone, and rare annelid, and other tribes that had harboured there in safety

---

49 Watney, H., "My Crass", *Hardwicke's Science Gossip* (1871), pp. 12–14, p. 13.

for centuries, put to the rout, while the young progeny, which were destined to supply future years with such animals, had been, in the wholesale destruction that had so recklessly been made, utterly annihilated.

This puts into perspective Lewes and Eliot's hammering and chiselling and crowbarring and levering. In fact, in Lewes we see two things: firstly, that he believed that the "lower animals" could not feel pain and so the odd hammer blow was not anything to them and, secondly, that he thought the things that were prised and cracked from the rocks grew back. One assumes that the many people who took his *Sea-Side Studies* with them on their own expeditions found no reason to dispute with the biographer of Goethe and thus an ecological catastrophe ensued that has yet to be repaired and probably never will be.

One cannot write about animals in the Victorian period without sooner or later mentioning the astonishing Frank Buckland. Although he had interests across the world of zoological curiosity – especially acclimatisation and what kind of eating various animals made – his most extensive work was in fish and, in addition to several other appointments, he was Inspector of Salmon Fisheries. Buckland's father, the equally astonishing Dean William Buckland, inculcated in Frank not only his general and eclectic interest in the animal world but also his love of fish:

> Pisciculture was a subject to which Buckland [William] devoted much attention. It was from his father that Frank Buckland must have inherited his taste for fish farming.[50]

Buckland was concerned firstly to ensure that the value of sea fish to the economy and as a foodstuff was fully exploited and also that the ecological conditions necessary for healthy freshwater fish were maintained. This latter concern was a focus of his role as an inspector and to promote the first he set up his Economic Fish Museum in one of the spaces in the new South Kensington Museum (now the Science

---

50   Gordon, E.O., *The Life and Correspondence of William Buckland* (London: John Murray, 1894), p. 186.

Museum). The museum borrowed from the name and organisation of Sir Henry De la Beche's Museum of Economic Geology (later Practical Geology) in Jermyn Street and Sir William Hooker's Museum of Economic Botany at Kew and so housed a great variety of things illustrative of the commercial and practical aspects of fish and fishing: a model of a salmon ladder, demonstrations of hatching tanks, all manner of equipment and paraphernalia, but above all the exquisite plaster casts of various species of fish that Buckland (who was a virtuoso caster and taxidermist) made and painted. Photographs of it show what looks not unlike a large and not particularly well-organised antique centre. This exhibition was set up at his own initiative and after some controversy with the management of the museum who took the reasonable view that their institution was for the display of artefacts drawn from its own holdings not private collections. But Buckland won the day and his display did apparently prove a popular part of the museum.

After Buckland's death, his message about the economic importance of the fishing industry continued to reverberate through various exhibitions and these could attract massive numbers. For example, the International Fisheries Exhibition at the Royal Horticultural Society's grounds in South Kensington ran between 12 May and 1 November 1883. During that time, it was visited by 2,689,092 people including roughly one million from London alone. It incorporated Buckland's Museum (which had by this time become part of the South Kensington collection under the terms of Buckland's will) and many of the displays were based on expanded versions of Buckland's exhibits. It also included the largest aquarium ever constructed with 39 tanks of various kinds. Many were filled with sea water shipped in from Brighton (so must have been a touch cloudy). The tanks included not only fish but many kinds of aquatic birds, reptiles and mammals. There was also an otter pond and a colony of beavers from Canada. It was a massive event that spawned a significant literature of its own. The daily average attendance was 18,545 with a maximum recorded of 25,000 and this average was maintained even on a Wednesday when a premium admission of half a crown (a day's wages for many workers) was charged to enable a more exclusive experience for those who could afford it.

There were many other such fisheries exhibitions – Norwich, Edinburgh and the Royal Aquarium at Westminster also mounted large events – but this was the biggest and most successful and the astonishing crowds it drew over many months show the extent to which the fascination with fish initiated by the modest display in Regent's Park 50 years earlier continued to grip the public. But now Victorians were encountering fish as an economic commodity and not just a scientific curiosity. Alas, after the closure of the South Kensington exhibition, Buckland's collection did not find its way back into the Natural History Museum but was left stranded:

> Since then [the end of the exhibition] the specimens have remained in a more or less dilapidated condition in the draughty and uninviting sheds which surround a portion of the grounds formerly in the possession of the Horticultural Society.
>
> (*The Times*, 22 February 1892)

Fortunately, they were eventually rescued and can now be seen in the Scottish Fisheries Museum at Anstruther, an establishment that Buckland himself might have founded and would have loved.

Buckland's role as inspector and his promotion of the economic value of fish led to two professionalising innovations. The first was the idea of a commercial fish farm. The second was the movement to establish a network of scientific marine research stations around the coast. It was a remarkable thing that, although fish created such interest and were increasingly seen as an important contributor to the nation's nutritional and economic health, surprisingly little was known about them. And although the aquarium movement had improved the position by which the pioneers of British ichthyology Ray and Willoughby were dependent on random finds in the bycatch of the Venetian fish market, there was still a way to go. As recently as 1828 the military surgeon John Vaughan Thompson – who, pursuing marine science as an amateur, made crucial discoveries about the development of crabs and barnacles and, using a dredge of his own design, more or less invented the study of plankton – had been so impoverished by the cost of publishing his own research that he had to apply for relief to the Linnean Society.[51]

4 She sells seashells

In 1872 HMS *Challenger* set out on a three-year voyage of deep-sea exploration under the scientific direction of Charles Wyville Thomson. He was an authority in deep-sea dredging (using equipment developed from Edward Forbes' dredge) and the *Challenger* expedition built on his previous experience of such journeys on HMS *Lightning* (1868) and HMS *Porcupine* (1869 and 1870). These voyages established the existence of life at very great depths (down to 2,435 fathoms). The *Challenger* was a revolutionary ship as it included not only a large scientific staff but also a marine science laboratory, an engraving of which shows a substantial room fully fitted with benches, microscopes and an extensive library. Thomson and his staff established that life could be found at all depths and in all seas whether these were tropical or Antarctic. His book *The Voyage of the Challenger* (1877) supplemented his previous *The Depths of the Sea* (1873) and set a whole new vision of the ocean world before the Victorian public. The theory of an "azoic zone" below which no life would be found was no longer tenable.[52] Creatures that were previously only known as fossils were found alive and well. The world, as it so often did in the Victorian period, suddenly opened up again and the meaning of the sea and the animals that lived in it continued to change.

Wealthy enthusiasts had various yachts fitted out as research vessels with every possible facility. For example, on 14 February 1891, *The Times* reported on Prince Albert I of Monaco's new yacht *The Princess Alice*. This had an onboard aquarium, fully equipped laboratories, a photographic darkroom, the most modern dredging equipment and comfortable cabins for the scientific staff. Prince Albert made important oceanographic discoveries and his various cruises were invaluable to marine science. He also set up an Oceanographic Museum in Monte Carlo. But it was clear that a network of nationally supported professional research stations would be necessary if the life cycles, breeding and migratory patterns of fish and other marine creatures

---

51 See Damkaer, D., "John Vaughan Thompson (1779–1847), Pioneer Planktonologist: A Life Renewed", *Journal of Crustacean Biology*, 36 (2016), pp. 256–62.
52 Yonge, C.M., *British Marine Life* (London: William Collins, 1944).

were to be properly and systematically understood in the interests of sustainable fishing and environmental monitoring.

Although the earliest of these was actually the "laboratoire des dunes" (1843), set up as a private venture in Ostend by the Belgian scientist Pierre-Joseph van Beneden, it is the Stazione Zoologica Anton Dohrn at Naples set up in 1872 (the same year as the Station Biologique de Roscoff in Brittany) by the scientist Anton Dohrn and still operating today that is usually credited as being the first modern marine sciences research institute.[53] It included an aquarium as well as research facilities and received large grants from Britain including, as previously noted, donations from the ZSL. It also received generous funding from the German government. In the absence of any similar facilities in the British Isles, it was felt that having a stake in the station at Naples would not only give privileged access to the scientific discoveries but also be a useful source of specimens for the Fish House. The Stazione had three income streams: public admission fees to its aquarium; sales of its three scientific journals; and the "bench system" by which institutions and individuals could rent a research space with all associated facilities for a year at a time. This model was innovative at the time but has become one of the standard ways of understanding and structuring research funding and it was a great success, especially since eminent scholars could also be invited to take a bench for the year.

It was not until June 1888 that the first British research station, the Plymouth Marine Biological Laboratory, opened and this was funded as a private–public partnership between the state and private philanthropy. It came into being out of a scientific dispute between

---

53 On the Dohrn Station, see Groeben, C., "The Stazione Zoological Anton Dohrn as a Place for the Circulation of Scientific Ideas: Vision and Management", in K.L. Anderson and C. Thierry (eds), Information for Responsible Fisheries: Libraries as Mediators, Proceedings of the 31st Conference (Fort Pierce: IAMSLIC, 2006), pp. 290–9 and for images of the facility as it looks today after extensive restoration, see Barling, S., "See Naples and Dive", *The World of Interiors* (July 2019), pp. 90–9. For detail on a specific British station which shows the line between scientific research and the keeping of an aquarium as a place for visitors, see Herdman, W.A., *Port Erin Biological Station: Guide to the Aquarium* (Liverpool: Liverpool Marine Biology Committee, 1906).

Huxley, who thought that fish stocks, especially of cod and herring, were so enormous that they could never be depleted by commercial fishing, and Edwin Lankester, who, correctly as it turned out, thought the opposite. The research institute was founded to monitor the progress of the unfolding ecological catastrophe of the decimation of Atlantic fish stocks and to understand what, if any, interventions could be made by using scientific evidence, especially of fish breeding cycles and migration patterns. It cost £13,000 to build and was projected to have an annual income of £900 (this compares with figures of £20,000 and £4,000 respectively for the institute in Naples). The laboratory included an aquarium, laboratory spaces for 24 scientists, a library, a photographic darkroom and living quarters. Resident staff included the director, two scientists, an engineer and his wife, a fisherman and two general purpose boys. It had a project to buy its own steamship and projected a funding stream from relays of fee-paying students based on the bench system pioneered in Naples.

The example of Plymouth led to other stations being established in an old barge *The Ark* at Granton and subsequently Millport (1884), Puffin Island (1889), St Andrews (1892), Port Erin (1892), Piel-in-Barrow (1896), Cullercoats (1897) and Aberdeen (1899).[54] These establishments helped to shift the balance of engagement in marine science firmly in favour of the professionals and the institutions but they build on a fascination with fish which grew out of the hybrid world of the professional-amateur aquarium. It has been well said that:

> The station movement also integrated ambitions, technologies and practices from a variety of other places of science. Some of these places were sites of scientific display, popularization and spectacle as much as they were venues of proper research. Particularly influential were the public aquarium and natural history museum.[55]

---

54 Kofoid, C.A., *The Biological Stations of Europe* (Washington: Government Printing Office, 1910) offers a comprehensive overview. See also de Bont, R., *Stations in the Field: A History of Place-based Animal Research 1870–1930* (Chicago: University of Chicago Press, 2015).

55 de Bont, R., *Stations in the Field: A History of Place-based Animal Research 1870–1930* (Chicago: University of Chicago Press, 2015), p. 39.

The marine research station in Jersey was not the one that William Saville-Kent hoped to set up but a private enterprise between William Hornell and Joseph Sinel. Sinel had previously been a taxidermist and dealer in preserved marine specimens. The station opened in 1893 and was run along the same lines as a state-funded institution, including the opportunity to study there as a visiting scientist. Hornell and Sinel also published *The Journal of Marine Zoology and Microscopy*, which offered serious scientific articles to a popular readership pursuing marine science at home. Between 1893 and 1897 subscribers could also receive sets of prepared slides with an accompanying booklet of photomicrographs sold as *Microscopical Studies of Marine Zoology*. These cost one guinea per year for the slides and booklets, or 8s for the booklets only and thus were within the range of many people, although requiring some saving at the lower end of the income scale. The station also sold live specimens of marine organisms. For example, £1 10s per year would get you a fortnightly consignment of starfish, worms and crustacea in a jar containing "between five to fifty or more forms" according to an advertisement in *Hardwicke's Science Gossip*. You could also buy individual prepared slides at between 1s 3d and 2s each or 35s for the set of 35 slides in a cloth-covered box, post free to the UK, Europe, Canada, India and Australia. The Jersey station was by no means the only such source of slides and specimens, and, although it closed in 1899, it was sufficiently successful to demonstrate that there was a significant amateur marine science community and also that, while the recreational home aquarium had become a rarity, the scientific home aquarium was still thriving.[56]

In fact, one might say that embedded in these stations was the learning from thousands of parlour aquariums and from the ladies who prowled the beaches looking for new species of seaweed and whose taxonomic endeavours remain today as their monument. At the same time, for the author of "An English Biological Station", an article published in *The Times* (31 March 1884), it was equally true that, while the better aquariums did offer some assistance to science (and he

---

56 See Stevenson, B., "Joseph Sinel, 1844–1929, James Hornell, 1865–1949", www.microscopist.net

especially mentions Brighton, the Crystal Palace and Birmingham), the fact was that:

> The great essential of all such institutions was and is that they should pay. They were regarded by their shareholders and managers as simply forming part of their by-show. At all events it finally became clear to biologists that science could expect little help from the ordinary aquarium which was no more than a handmaid to amuse the public.

As we shall see in a subsequent chapter, the public aquarium movement was largely founded on the hopes of a good dividend to the shareholders, and that the entertainments available could sometimes result in the display of fish played a very minor role.

Buckland's legacy also extended to the development of commercial fish farms and his various experiments with breeding tanks for trout and salmon at South Kensington, the Zoological Gardens and the Crystal Palace (and also those of his rival Francis Francis at Twickenham) in particular led, in 1875, to the establishment of the first commercial fish farm in the world at Howietoun in Scotland. Concarneau had been founded in France in 1859 but, although this was originally a breeding station, especially for crustacea and flatfish (and then developed into a more general marine research institute), it was a scientific rather than a commercial concern. The Howietoun farm was set up by Sir James Maitland, who had been experimenting with fish breeding techniques since at least 1873. The importance of the farm was twofold. Firstly, Maitland was a keen scientist and made meticulous records of the progress of his various experiments with breeding. These he wrote up in his *History of Howietoun* (1887).[57] Secondly, Maitland was concerned with the acclimatisation of trout and salmon in the colonies and developed transportation systems that enabled viable ova to be transported transhemispherically. In this way the rivers of New Zealand were populated with acclimatised Scottish fish and, as we shall see in a subsequent chapter, these experiments were

---

57   Maitland, J.R.G., *History of Howietoun* (Stirling: Howietoun Fishery, 1887).

influenced by earlier fish transportation which eventually led to fish farming and new populations of fish in Tasmania.

The salmon and trout that supplemented the wild populations were, of course, the prey of human anglers. The final aspect of the Victorian encounter with fish in managed nature is its double manifestation in the boom in fishing as a hobby. Double because as the period wore on, a distinction emerged between game fishing (for salmon and trout) and coarse fishing (for other freshwater fish). The distinction, however, was not just a matter of target species; it comprised also a world of differences: class, gender, technology and environmental impact were all in play as the Victorians invented for themselves a whole new domain of social stratification that set nineteenth-century fishing quite apart from the jolly camaraderie of Walton and Cotton 200 years earlier.

The distinction is easiest to understand if we look at the costs involved in setting up as a game fisherman (although many women took part in the sport) as opposed to those involved in setting up as a coarse fisherman. In coarse fishing, women were almost unknown in the later Victorian period, although before the development of the game/coarse fishing distinction there is plenty of evidence that women enjoyed everyday fishing.[58] Yet as early as 1828 when Sarah Bowdich Lee began the process of collecting fish to paint for her *Fresh-water Fishes of Great Britain* she informed the reader:

> It had never been my intention to touch upon the manner of catching the fish that I have delineated, for that demands an experience and skill that a female cannot be expected to possess.[59]

Instead, she employed men to catch the fish for her while she sat on the bank ready to sketch the specimens that were pulled out of the

---

58 Although D. McMurray's *A Rod of Her Own: Women and Angling in Victorian North America* (M.A. thesis, University of Calgary, 2007) is concerned with North America, it has contextualising material relating to the early history of women anglers in England.

59 Quoted in Gates, B., *Kindred Nature* (Chicago: University of Chicago Press, 1998), p. 78.

4 She sells seashells

water. When she set up at Henley-on-Thames, the locals caught fish for her in exchange for a sight of the finished watercolours (which were experimentally embellished with silver and gold leaf to reproduce the effect of shimmering scales). This disclaimer on her part is an astonishing one when you consider that she was the first English woman to do zoological fieldwork in West Africa, that she became a close associate of Cuvier and that her work was admired by Humboldt. She was clearly capable of picking up a fishing rod, so I think these words show her care to establish a persona that would encourage subscribers to buy what was a de luxe product.

Let us assume a clergyman of comfortable background in possession of a good living – perhaps along the lines of Mr Collins' £500 in *Pride and Prejudice* or Mr. Bolton's £700 in *Mansfield Park* – somewhere in Devon in, say, 1875. He wishes to set up as a game fisherman and so he buys a copy of the Plymouth company J.N. Hearder & Son's *Guide to Sea Fishing and the Rivers of South Devon: and Descriptive Catalogue of the Prize River Fishing Tackle, Cricket, Archery, Croquet, Umbrellas, Parasols etc.* He pores over it with glee and decides to invest in a full set of kit for both salmon and trout. He is not a landed aristocrat, so there is a certain degree of interoperability between his tools, but he buys the best quality he can afford. This is what his equipment looks like:

1 fly rod. He could pay anything between 8s 6d and four guineas for this, but he settles on a greenheart rod at £1 4s.
1 salmon rod. £3. He needs both, as the lighter trout rod might not be strong enough for salmon and the heavier salmon rod would not cast delicately enough for trout.
A reel for each rod, 12s for salmon, 6s for trout.
20 yards of silk and horsehair line for trout, 6s 8d, and 30 yards for salmon, 10s.
Gut for trout, 30 yards at 7s 6d.
Gut for salmon, 30 yards at £1 2s 6d.
Two hanks of fine fly gut, 10s.
3 dozen trout flies, 4s 6d.
3 dozen sea trout flies, 7s 6d.

2 dozen salmon flies (these vary wildly in price with the most costly being 7s each). Our Reverend gentleman is more modest and intends in the fullness of time to tie his own flies, so he compromises, and the total cost is £1 4s.
Artificial lures or "minnows". A dozen for trout, 12s, and a dozen for salmon, 16s.
3 "Devon" minnows; these were expensive at 3s 6d each but he has heard that they are supremely effective, so he buys three, 10s 6d.
3 "phantoms", another top of the range lure for salmon, 12s.
Worm tackle, 2s.
A landing net big enough for salmon, 2s 6d.
A telescopic handle, 6s.
Leads, 1s.
Swivels, 1s 6d.
Worming hooks, 1s 6d.
A fly book, 7s 6d (the most expensive but he wants to be known for his flies).
A spring balance for weighing the catch, 9s.
Wax to dress the line, 2d.
A fishing basket, 5s.

That all comes to £13 11s 4d, which is a very significant outlay when set against the kind of wages quoted in a previous chapter in the context of the price of home aquariums. And this doesn't include the cost of his fly tying.

The practice of fly tying became more and more elaborate during the nineteenth century and the cost of the rare feathers recommended for the really "killing" flies was another consideration that separated the gentleman and lady anglers of the trout and salmon streams from their humbler compatriots on the river and canal banks. The first fly tying book, William Blacker's *Art of Fly Making*, appeared in 1842 and required that the would-be aficionado supplied himself with the plumes of rare birds such as the toucan, the South American cock of the rock, the Indian blue kingfisher, the resplendent quetzal, the bird of paradise and Amazonian macaws. There was no particular evidence that these flies were more effective than the older confections of dyed chicken feathers but individual anglers and clubs started to develop their own patterns and there was, increasingly, a display of conspicuous consumption in rarer and rarer feathers that, of course, helped further to distance the practitioners of "game" fishing from the rest of the fishing fraternity.[60]

By 1895, George Mortimer Kelson's *The Salmon Fly* had taken the obsession with rare feathers to new levels. In addition to promulgating a whole science of fly tying, his recommendations for feathers included those of the banded chatterer and the golden bird of paradise (just one of which cost £10). Between Blacker and Kelson came works by Hewett Wheatley (1849), Edward Fitzgibbon (1850), Henry Newland (1851), Thomas Stoddart (1853), Francis Francis (1867) and Fraser Sandeman (1894). All these contributed to a significant industry that ran parallel to the sport of game fishing. And while in the later nineteenth century millinery was starting to draw criticism for its profligate use of feathers, especially those of rare or endangered species, the anglers' assault on fragile bird populations was carried out slightly less visibly – indeed, underwater. And if the fish farming which helped stock the game fishing rivers added to the fish population (not always with happy environmental results), the effort to remove those fish from those rivers also removed a good many birds and the populations of some (such as the banded chatterer) have still not recovered.[61] So, the Victorian encounter with fish could have ramifications far beyond the aquarium or riverbank. And our Devonshire vicar could easily add £20 a year to the cost of his hobby if he was in earnest about fly tying.

If game fishing was insulated by price as the province of the well-to-do, coarse fishing involved more people and less money. In his *A Book of Angling* (1876), Francis Francis recommended that the coarse fisherman required a bamboo rod, a reel, at least two yards of gut, a cork float and a few dozen hooks of different sizes. He does not give prices

---

60   On the history of exotic feathers in fly tying and much more, see Johnson, K.W., *The Feather Thief* (London: Hutchinson, 2018).
61   See Boase, T., *Mrs Pankhurst's Purple Feather* (London: Aurum Press, 2018). A campaign by which ladies were urged to write letters to any of their friends they saw wearing feathers started the Royal Society for the Protection of Birds through an amalgamation with the Fur, Fin and Feather Folk in 1889; see Moss, S., *Birds Britannia* (London: HarperCollins Publishers, 2011).

for these but a similar list in J.J. Martin's *The Nottingham Style of Float Fishing and Spinning* (1885) does. Martin specifically says that:

> The extent of the pocket of the working man has been constantly before me when describing his outfit and there is nothing mentioned that cannot be bought or made cheaply.[62]

He recommends:

> A rod, 6s 6d.
> A reel, 5s (but can be much less).
> Line, 1s 6d.
> Four spools of silk, 1d each.
> A home-made tackle box and float, hooks would be extra 1s at most.

This all comes to 14s 10d , which is still a good chunk of a week's wage but looks very affordable if spread over a few weeks and, in addition, a basic roach pole or simple bamboo rod could be had for a couple of shillings so the main expense in this list could be considerably reduced.

This kind of cost was plainly at a level that enabled mass participation and, by 1890, London could boast 620 angling clubs while Sheffield had 200 with a combined membership of 21,000. These clubs included few women and there were probably two main reasons for this. The first was that in the early days, at any rate, fishing clubs appear to have been places where beer and strong language could flow and that did not encourage women to join even if they were allowed to; secondly, there may well have been the beginning of a specific gendering of time that promoted the separation of men and women into different spheres: the women into the domestic and the men into the outdoor. Even so, this separation does not seem to apply to natural history and cycling clubs, which were both outdoor and with numerous women members and included many who were predominantly working class especially

---

62   Martin, J.W., *The Nottingham Style of Float Fishing and Spinning* (London: Sampson, Lowe, Marston & Rivington, 1885), p. vi.

4 She sells seashells

in the north of England. So, this exclusion of women may also be about the identification of coarse fishing as a specifically male sport.[63]

Marine science and the sea and the human encounter with fish permeated Victorian experience. The public and private institutional domains intersect here with managed nature and, indeed, the perceptions of fish constructed in those domains sculpt and determine the way fish are perceived in "nature". When a Sheffield coalminer pulled a roach out of a canal, he was not just catching a fish: he was encountering a complex and over-determined cultural product and participating in a richly over-determined chain of signification and value.

---

63 Locker, A., "The Social History of Coarse Angling in England AD 1750-1950", *Anthropozoologica*, 49 (2014), pp. 99-107 and Locker, A., *Freshwater Fish in England* (Oxford: Oxbow Books, 2018).

Plate 1 The pike in the Regent's Park Fish House photographed by the Comte de Montízon and admired by William Alford Lloyd before the building opened.

Plate 2 The exterior of the Regent's Park Fish House showing what a modest building it was and how it reflected contemporary enthusiasm for glass buildings with horticultural models.

Plate 3 The interior of the Fish House. It is, perhaps, hard for us now to understand just how revolutionary the experience afforded by these tanks was to the Victorian public.

Plate 4 The exterior of the Dublin Aquatic Vivarium. This photograph is undated but the skylights at the end of the building show that it must have been taken after 1887.

Plate 5 "The Pets of the Aquarium", an illustration from George Kearley's book *Links in the Chain; or, Popular Chapters on the Curiosities of Animal Life* (1862). This shows the way in which the home aquarium was readily associated with the women and girls of the house, although in practice the great majority of commentators who talked about keeping their own aquariums were men.

Plate 6 "Goldfish" by Carlton Smith (1895). This watercolour shows the most basic sort of aquarium in a humble domestic interior. As there is no sign of any plants to aerate the water the goldfish will die within a day or so.

Plate 7 "The Seaside in Autumn" by Robert Loudan Sr. (1858). This postdates Frith's "Ramsgate Sands" by about seven years and it is interesting to note how, in that intervening time, the beach has been populated by the signs of marine science. We see ladies and a gentleman with hammers and other tools, a girl collecting seaweed, a boy capturing a crab, and an artist sketching with scientific handbooks nearby.

Plate 8 "A Contrast" by Abraham Solomon (1855). This scene shows the real conflict in the Victorian presentation of the beach and the seaside experience. On the one hand it is a place of science and genteel recreation, on the other it is a workplace where the classes are doomed to come together in potentially difficult encounters.

Plate 9 "Seaside Sirens" by Jules Pelcoq, under the name of Jules Pelcor (1855). Here is the feminised scientific seaside at full throttle. None of the ladies are dressed as Margaret Gatty or G.H. Lewes would recommend. One is busily using a crowbar to disturb a rock and, no doubt, kill the animals underneath. It is hard to know whether the two gentlemen who have come across this scene are shocked or excited.

Plate 10 A small Victorian collection of seashells belonging to the author. The labelling shows they were gathered in Australia (on Lord Howe Island), Trinidad, Teneriffe, Argentina and South Africa.

Plate 11 "The Brighton Aquarium" (1872). Notice how affluent the clientele appear and not at all like the mass of Londoners who were enticed down on cheap day trips. The young lady to the left of the central column is holding what I assume to be Saville-Kent's Guide. By the standards of other aquarium guides this was a cheap production as it had no engravings, but this may have been a function of the rush to get it out rather than a deliberate editorial decision.

Plate 12 "The Crystal Palace Aquarium" (1871). The tanks are a combination of Regent's Park-style table tanks designed to be looked down upon and large wall-mounted tanks to stroll past. Notice the octopus dimly visible in the wall tank on the right.

Plate 13 "The Aquarium at the Southport Pavilion" (1874). All the tanks are wall-mounted and the patrons appear to be well-to-do.

Plate 14 "The Yarmouth Aquarium and Winter Garden" (1876). This image illustrates the extraordinary scale of many of the Victorian aquarium projects and shows why they usually failed to thrive. Notice the Kew-like wing housing the Winter Garden.

Plate 15 "L'aquarium d'eau douce, dans le parc du Trocadéro" (1878) at the Paris Exhibition. Here is grotto architecture in a highly developed form. Notice that the tanks are set deep into the cave-like niches of the artificial rock.

5
The public aquariums 1

In this and the following chapter, the expansion of the aquarium movement into large-scale public aquariums will be considered in some detail. The focus will be less on the fish and more on the specific institutions and the social and economic conditions that made the building of large aquariums such a feature of English life in the 1870s. The boom in the development of public aquariums mirrors the boom in zoo building 20 years earlier and, for the most part, they incorporated into their missions a dual responsibility: to provide spaces for scientific education and to offer opportunities for the respectable use of leisure time (especially for the lower classes). As will be seen, most of the new institutions failed for the same reason that most of the municipal zoos failed – which was that a collection of fish is not, in itself, a sufficient attraction to guarantee the repeat business required for long-term financial sustainability and for a competitive return to shareholders. However, the word "aquarium" seemed to cast an almost magical spell over would-be investors and the detailed discussions in shareholders meetings that will be set out below show how, while some did have a realistic idea of the worth of their investment, others were overly optimistic and naïve as to the potential of the businesses into which they had put their money.

A significant difference between the zoo-building boom and the aquarium-building boom was that, while most of the zoos were funded by public money through local authorities (Regent's Park was an exception, being the property of a private learned society as was the most successful of the new zoos, Belle Vue in Manchester, which was a private business and had many attractions alongside its menagerie) most of the aquariums were funded by locally formed private companies. So, the failure of a zoo was not usually attended by the acrimony that followed the failure of an aquarium. The main determinant of this difference was a change in the legal framework for setting up companies and raising capital that made it possible for speculative ventures to be funded by the selling of fixed numbers of shares at a fixed price. Without this mechanism I doubt that the aquarium boom would have happened as many local authorities may have remembered their zoo venture and would have had no desire to repeat the mistake.

The boom in public aquariums also followed on after the craze for domestic aquariums. As we have seen, the vast majority of these failed so the question might legitimately be asked why, given this experience, did shareholders (many of whom must have had personal experience, direct or indirect, of failed domestic ventures) in aquarium companies believe that they would do any better? The answer to this lies in the success of a new wave of aquariums based on novel technologies in Europe and especially in France and Germany. These gave confidence that large-scale aquariums were viable and, for the most part, they were designed and constructed using English expertise. As *The Times* (18 October 1872) put it:

> This construction of aquaria we are entitled to boast is eminently and entirely an English feat of scientific industry, for all the successful aquaria on the Continent have been, and are being, constructed from English plans and under English supervision.

These new European aquariums will not be looked at in detail here and only mentioned where they help to explain the development of aquariums in England.

Given all this, the shareholder may legitimately have thought that to build a successful aquarium it was simply a question of harnessing that English expertise at the right price. However, as will be seen, the

5 The public aquariums 1

difficulties of building and sustaining a healthy aquarium, especially a marine aquarium, using steam-driven technology are not negligible and even large and well-funded establishments would, on occasion, falter. In addition, unlike a zoo which, at its most basic, simply requires fencing off a bit of public land and erecting some enclosures, an aquarium requires a large building on privately purchased land equipped with highly specialised technology for lighting and temperature control, pumps, boilers, yards of pipework, an aeration or circulation system potentially requiring huge storage tanks and, of course, large tanks made of plate glass and filled with sea water. And these did not, if properly constructed and arranged, come cheap.

Finally, the boom in aquariums was driven by two large institutions: Crystal Palace and Brighton. These provided contrasting but initially successful models for such public establishments. In particular, the shareholders of both institutions enjoyed very significant dividends in the first few years of operation and these, rather than the long-term fundamentals of the business, were what often attracted investors in the local aquarium companies. In fact, the economic viability of aquariums was never especially promising from the early days, not least because the costs of running them were so high. Lloyd himself believed that no aquarium that was only an aquarium could pay its way unless the initial capital investment was no more than £3,000 and the annual revenue expenditure no more than £500. And you couldn't get much of an aquarium for that. As *The Spectator* (8 January 1875) put it in an article entitled "Fish Shows as Pecuniary Speculation" (which also quoted the opinion of Lloyd cited above):

> A really great Aquarium, "to pay," must be something more than an aquarium; or, if not that in itself, it would require to form a portion of a greater exhibition such as that of the Crystal Palace, where the Aquarium is a parasite of the larger building.

This is precisely the lesson that should have been learned from the zoo-building boom. But it was not. And even the highly popular European aquariums were not necessarily good businesses. For example, *The Times* (23 January 1872) reported that the revenue of the Berlin Aquarium in 1870 was 67 per cent lower than the revenue in its opening year of 1869, presumably because in that first year everyone who wanted to see it saw it and, having seen it, didn't come back. *The Spectator* article quoted above points out that, although Lloyd's Jardin d'Acclimatation aquarium in Paris had 450,000 visitors annually, it had never paid a dividend and that the aquarium at Hamburg, which was profitable, prospered on the basis of astonishing visitor numbers (up to a quarter of a million a week – equivalent to the entire population of the city) to the gardens where it was situated, a modest capital outlay of only £3,000 and iron discipline in keeping running costs at only £600 per annum. It was this case that Lloyd considered proved the correctness of his view "that no fish-show will ever pay for itself unless the expenditure be restricted to the sum named".

It is now time to consider the pioneering institutions.

The Crystal Palace and Brighton aquariums

The small scale of the Regent's Park Fish House and its increasingly obvious shortcomings combined with the waxing and waning of the parlour aquarium craze meant that for the remainder of the 1850s and 1860s Britain was without any new aquariums. However, on the continent of Europe and, to an extent, in the United States, things were quite different and the lessons learned in London and Dublin (and perhaps, it should be added, in the aquariums set up by P.T. Barnum in New York in 1857 and later in Boston in 1860) were being put to good use.[1] It is also worth mentioning the small aquarium set up in the Smithsonian Museum by William Stimpson (whom we have met before as a pioneer of the home aquarium). The *Daily National Intelligencer* (2 November 1857) tells us that this had failed before "for want of a supply of sea water" but was now successful. Other sources tell us that the tank was five feet by three feet and 18 inches high and contained around 300 animals drawn from 38 different species and that various donations of fish and other creatures were

---

1   See Saxon, A.H., "P.T. Barnum and the The American Museum", *Wilson Quarterly*, 13 (1989), pp. 130–9.

5 The public aquariums 1

made in 1857 and 1858. This was a small affair but it was the first genuinely scientific public aquarium in the United States and perhaps bears out Stimpson's claim (early failures notwithstanding) that he had solved many of the problems of keeping a marine tank healthy independently and as early as the English aquarists.[2] Large aquariums of a scale and kind not dreamed of by the ZSL or by Gosse or Lloyd when he first saw the pike swimming were developed in France and Germany. These not only had larger tanks and more sophisticated filtration and aeration systems than had previously been possible but also successfully bridged the gap between the aquarium as a place of scientific discovery and education and the aquarium as a place of entertainment and leisure. And, although the focus of this book is exclusively on aquariums in Britain, it is probably easier to understand how the next wave of aquarium building happened there if we understand a little of the nature of the new public facilities in Europe.

Large public aquariums were opened in Paris (1861), Hamburg (1864), Hanover (1866), a second in Paris (1866), Berlin (1869), Munster (1875), New York (1876), Frankfurt (1877), Leipzig (1878), Zurich (1885) and Rome (1887). The aquarium at Sebastopol, which opened in 1897, specialised only in the fish found in the Black Sea and had developed as an adjunct to the marine research station that had been established there some 20 years earlier. And there were many other smaller establishments: for example in Boston (1859) and Vienna (1860).[3] In 1882, a major new aquarium was opened at the important Artis Zoo in Amsterdam and this had a number of distinctive features: it was the first of the Artis buildings to include a scientific infrastructure in the form of a lecture theatre and laboratories, and a qualified scientist, Coenraad Kerbert, as its designer and curator. The building had a separate entrance for members of the Koninklijk Zoologisch Genootschap *Natura Artis Magistra* (the Royal Zoological Society *Nature is the Teacher of Art*) which, as with the RZSL in London, owned

---

2   Vasile, R.S., *William Stimpson and the Golden Age of American Natural History* (Ithaca, NY: Cornell University Press, 2018), p. 208.
3   Ryan, J., *The Forgotten Aquariums of Boston* (Pascoag, RI: Finley Aquatic Books, 2002).

and ran the gardens.[4] In the first year, visitors to the aquarium who also wandered round the zoo could have seen the last-known quagga, which died in captivity there in 1883.

Outside the Euro-American world, the Khedive Ismail Pasha's aquarium at Gezira was first set out in 1874 but not opened until 1902 when it was reconstructed under English management. But, nevertheless, the taste for menageries had developed among the Egyptian ruling class before Egypt came under British rule. As Gustave Loisel wrote:

> Avant l'occupation Européenne le Khédive Ismail Pasha avait developpé le gout des menageries.
>
> [The Khedive Ismail Pasha had developed the taste for menageries before the European occupation.][5]

And a letter from a Mr John Stuart published in the *Friend of India and Statesman* (12 December 1881) made an impassioned plea for an aquarium in Calcutta. Stuart feared that if this did not happen, Bombay would get there first. He was well aware that, in England, aquariums were now failing as businesses but sensibly pointed out that there is "a difference between having too many aquaria and having none at all". As it happened, India would have to wait until 1909 for its first aquarium. This was at Madras (now Chennai) and opened in 1909 with an impressive selection of fish and a high-quality filtration system. The Madras government published *A Guide to the Marine Aquarium* in 1912, which gave a detailed account both of the collection and the technical specifications of the purification system.[6] Although a zoo

---

4   On this zoo, see Mehos, D., *Science and Culture for Members Only: The Amsterdam Zoo Artis in the Nineteenth Century* (Amsterdam: Amsterdam University Press, 2006).
5   Loisel, G.A.A., *Histoire des Ménageries*, 3 vols (Paris: Octave Doin et Fils, 1912), vol. 3, p. 28.
6   The history of zoos and animal displays in India is a complex and fascinating topic awaiting proper research. For a brief description of the Madras Aquarium, see Anon., "The Marine Aquarium Madras", *Nature*, 82 (1910), pp. 411–12.

5 The public aquariums 1

opened in the Botanic Gardens in Singapore in 1875 and ran relatively successfully until 1901, there was, surprisingly, never an attempt to add an aquarium.[7]

However, for the purposes of this study only two of these establishments need to concern us and they are the aquarium in the Jardin d'Acclimatation in Paris, which opened in 1861, and the aquarium in Hamburg, which opened in 1864. The reason is that they were both designed and initially managed by William Alford Lloyd. In the case of Hamburg, he came on the recommendation of Richard Owen: the second time that this eminent scientist had rendered him crucial assistance, which contrasts with the rather forbidding and unpleasant picture we have of him as the legacy of his disputes with Darwin. It was the things Lloyd was able to put into practice in these two European institutions that enabled him to establish his reputation as the foremost aquarist in Europe (or the world) and to set up in the Crystal Palace on his return to England what would prove to be the most advanced and successful (from a scientific point of view) aquarium yet seen.

The original plan of la Société Zoologique d'Acclimatation was to have the aquarium set up by J.D. Mitchell whom the Société had appointed as the new manager of its zoo in the Bois de Boulogne. This was natural enough as Mitchell's experience of London Zoo made him one of the most senior zoo professionals in the world and, in addition, it was he who had established the Fish House. The French zoologists would have known that the Fish House was by no means a perfect design. Louis Figuier said of it in an article on the Jardin d'Acclimatation aquarium in *L'Année Scientifique et Industrielle* (8, 1864):

Cet aquarium était pourtant loin d'être parfait.

[This aquarium was, however, far from being perfect.]

But the Société would have known that Mitchell had the business acumen to develop a facility that would attract people and a network

---

7   Barnard, T., *Nature's Colony* (Singapore: National University of Singapore Press, 2016).

that meant that he could call on the foremost scientists of the age for advice and direction. Unfortunately, Mitchell passed away not long after accepting the post in Paris. But he had clearly already been considering the needs of a new aquarium and I suspect that it was on his advice that the Société called in William Alford Lloyd who set about trying out his theories about water maintenance on a grand scale. As Guillaume le Gell has pointed out, the Société was ambitious and had aspirations to build something which, like the Fish House, would set a new direction for marine science and aquariums: "la Société ... fait appel à Alfred Lloyd ... pour imaginer le plus grand aquarium jamais edifié" (the society called on Alfred Lloyd to devise the biggest aquarium ever built).[8] It was "un aquarium ou fût appliqué, pour la première fois, le système à circulation d'eau continue de W. A. Lloyd" (an aquarium where for the first time W.A. Lloyd's system of continuous circulation of water was applied) and "der erster grosser aquarium dieser Art" (the first large aquarium of this kind).[9] And the early accounts of the aquarium – by Lloyd himself in the *Bulletin de la Société Imperiale d'Acclimatation* (9, 1862) and by the director of the Société, Rufz de Lavison, in the same journal (10, 1863) – spend almost as much time describing the engineering of the pumping system, the volumes of water held in the tanks and reservoirs and the extraordinary expanses of glass as they do the fish.

In addition to installing the pumping and aeration systems that would enable sea-water tanks to be kept fresh without constant replenishment, Lloyd incorporated into the design an innovation which would not only revolutionise the contemporary experience of visiting an aquarium but also set the pattern for almost every aquarium ever since. In his own warehouse and in the Fish House, the tanks were set out on tables, and, although you could look in at the side in the larger tanks of the Fish House, the experience was essentially that of a home aquarium replicated many times and on a bigger scale. In the

---

8 Le Gell, G., "Dioramas Aquatiques: Théophile Gautier Visite l'Aquarium du Jardin d'Acclimatation", *Culture et Musées*, 32 (2018), pp. 81–106, p. 85.
9 Lindheimer, O., "Aquarium" in *Gebaude für Sammlungen und Austellung (Buildings for Collections and Exhibitions)* (Darmstadt: A. Bergstrasser, 1893), p. 455.

Paris aquarium the tanks were set into the walls so that the viewer saw them like moving pictures and the experience of walking from tank to tank became somewhat like the experience of walking from painting to painting in a gallery. This entirely changed the way people saw fish and marked an important transition. In Regent's Park and Dublin (and indeed in Lloyd's own aquarium warehouse that was really an aquarium in its own right), the perspective was mostly, but not exclusively, downwards as the tanks were set up on tables. You could see into the sides of the tanks, but you more readily and more often looked down into them. At this point, anemones, crustaceans, worms and starfish were much more likely to be seen than large free-swimming fish. These were the inhabitants of the beach rockpools in which the Victorians first encountered marine science. So, the transition from the downward to the lateral view also marked the transition from the littoral marine science of Gosse, which was based on the rockpool, to a new science based on the sea itself where the aquarium was not so much a proxy for the beach and the seaside but for an impossible underwater exploration. The *Illustrated London News* (15 February 1890) picked this up in an article on a descent in an early submarine:

> We lived in a sort of vivarium, with an aquarium outside it, and just as – when we visited the Zoological Gardens – we were wont to stare at the jack and perch, so now did pike and perch, or their marine equivalents, come to stare at us, and wonder how we liked it.

This shift in perspective was emphasised in the extensive use of grotto architecture (of which Lloyd deeply disapproved) in the design of several large European aquariums later in the century, especially the Paris aquarium in Boulevard Montmartre, the aquarium set up in that city for the Exposition Universelle of 1867, and the Berlin Aquarium. In Paris the tanks also included live mermaids projected into the water using the theatrical illusion known as Pepper's Ghost. The Brussels aquarium, which was described in *The Telegraph* (16 August 1889) as "the most complete and splendid achievement of its kind", was also "a huge grotto roofed with imitation stalactites and divided into some half a dozen caverns".[10]

After his tenure in Paris was completed, Lloyd moved on to design another large aquarium in Hamburg and this opened in 1864. Again, Lloyd installed huge reservoirs to enable the constant circulation that was his trademark. Loisel, writing in 1912, was able to report that:

L'eau de mer n'a été renouvelée que deux fois depuis 1864; en 1898 et 1904.

[The sea water has only been renewed twice since 1864: in 1898 and 1904.][11]

So, the Hamburg aquarium, which the popular German magazine *Die Gartenlaube* (1863) described as "ein wahrer Triumph der Wissenschaft und Schönheit" (a true triumph of science and beauty), proved beyond doubt what Lloyd had always argued: that if you kept sea water in constant circulation you did not have to go to the considerable expense of refreshing it frequently.

The experience of building aquariums larger than anything that had ever been seen before in Paris and Hamburg meant that when an aquarium of similar scale was mooted for the Crystal Palace, Lloyd was the only realistic choice for designer and director.

After the Great Exhibition closed at the end of 1851 there were several ideas concerning the future of the Crystal Palace. In the end, the building was sold to the Crystal Palace Company, dismantled and re-erected to a different and much extended design at Sydenham and opened, by Queen Victoria, in 1854. There it became a site of commercial entertainment and recreation combining art exhibitions and other interesting displays, such as the extensive and spectacular Egyptian Court, with concerts and restaurants and the chance simply to wander in an extensive park containing many attractions such as a series of life-size sculptures of dinosaurs by Benjamin Waterhouse Hawkins. The Palace was connected to London by a rail link (there were two purpose-built Crystal Palace railway stations as well as the

---

10 On grotto style, see Adamowsky, N., *The Mysterious Science of the Sea, 1775–1943* (London: Routledge, 2016), pp. 109–10.
11 Loisel, G.A.A., *Histoire des Ménageries*, 3 vols (Paris: Octave Doin et Fils, 1912), vol. 3, p. 265.

existing Penge West station) which improved over the years. In its new manifestation, the Palace initially proved an immensely popular attraction.

In 1866 there was a disaster. A major fire destroyed the north end of the building. By this time the novelty of the Palace had worn off and while there was still plenty of custom there was not sufficient money to rebuild and restore what had been lost. So, when a new company, the Crystal Palace Aquarium Company, proposed building an aquarium on the site of the fire by utilising the extensive underground spaces that had survived, the proprietors and shareholders of the enterprise were only too pleased to accept their offer. Not least because the aquarium company would bear the costs of construction and, once the capital expenditure had been paid off through the revenue raised from the entry fee and associated merchandise (such as Lloyd's extensive and authoritative *Handbook to the Crystal Palace Aquarium*), there would be a new revenue stream into the parent company. The new company had raised its capital by the kind of joint stock share issue now possible under recently enacted legislation and would pay an annual dividend to the investors who were invited, as recorded by the *Morning Post* (27 May 1870), to buy £2 shares in order to build an initial capital fund of £10,000. The Crystal Palace Company rented the space to the aquarium at a nominal £50 p.a. (*The Times*, 3 June 1870), and itself invested £3,200 in a capital project to provide a covered walk from the Orangery in the original building to the new aquarium (*Morning Post*, 12 December 1870). The model adopted for the aquarium was seen as sufficiently attractive for another company to be set up to develop a museum of taxidermy within the main building of the Crystal Palace on what seem like the generous terms of rent at £100 per annum plus 10 per cent of all profits from the extra admission cost (*Morning Post*, 23 December 1873).

Under Lloyd's direction, an extensive pumping and reservoir system was installed that featured the "belt and braces" safety feature of having everything in duplicate – two boilers, two steam pumps, two sets of pipes (made of vulcanite because this is not corroded by salt water) etc. – to ensure that, whatever reasonably foreseeable breakdown occurred, the system would always be able to maintain the freshness of the sea water in the 60 tanks. Thirty-eight of these were on permanent display while 22 kept reserve stocks "to enable the company to purchase specimens from the various depots at such times as they may be most plentiful and cheap" (*The Standard*, 25 March 1871).

The vast quantity of water required (20,000 gallons for the tanks and 100,000 for the reservoir) was shipped in by rail from Brighton by the furniture removalist William Hudson who transferred the massive casks required by taking his horse-drawn vans directly onto the flat cars of the London to Brighton railway company, which offered "an almost nominal rate" (*The Times*, 23 January 1872) for the transportation (the original proposers and major shareholders of the original Crystal Palace Company were directors of this railway company that also built and owned the two Crystal Palace railway stations – an early example of a vertically integrated business model). Hudson was a good choice as he already had experience of shipping sea water to London to feed the sea water baths, which were an expensive treat – in 1844 the Lothbury Baths were charging 3s 6d for a cold bath and 7s 6d for a hot one – for the well-to-do (including Queen Victoria herself). The glass – the thickest and largest plates ever made – was produced by Alfred Goslett of Soho Square who had been an important supplier to the market for new glass products for greenhouses and conservatories since at least the late 1840s. It should be noted that, at this point and for several years after, the Crystal Palace Aquarium was a sea water only establishment – which drew some criticism:

> An aquarium without a fresh-water division is only half an aquarium.
>
> (*Morning Post*, 26 March 1875)

But this deficit does not appear to have unduly affected the initial attractiveness of the facility to the paying public. The aquarium was also lit by the still quite new electric light, as well as having a gas lighting system specifically to ensure the best visibility possible within the tanks housing the nocturnal animals. This was so effective that when the Prince and Princess of Wales visited, the *Daily News* (19 July 1872) reported that:

> The illumination extended to the tank so completely that every grain of sand was plainly visible on the turning up of the lights.

5 The public aquariums 1

Living, as we do, in the constant glare of light pollution, we forget how astonishing bright lights and the effects of artificial lights appeared to the Victorians.[12]

The first task for Lloyd was to purify and stabilise the sea water. This took some time and delayed the opening. A meeting of the shareholders was told that "Mr Lloyd had nearly overcome the obstinacy of the sea water" (*Morning Post*, 28 June 1871) but, of course, once the water was of sufficient quality Lloyd's tried-and-tested circulation system would keep it pure almost indefinitely. Lloyd was also, given his awareness of failures of temperature control in the Fish House, mindful of the crucial importance to the health of the tanks and the fish of ensuring that heat was rigorously moderated. In the summer the maximum temperature in the aquarium was 68°F and in the tanks 63°F and in the winter the gas heating system maintained the aquarium at 60–65°F and the water in the tanks at 55°F. This enabled Lloyd to keep his animals healthy and his tanks well stocked – one of his criticisms of rival aquariums based on the aeration system was that they could only support a few fish in the same volume of water that supported many in the unfiltered but permanently circulated water of his system. In his article "Aquaria" (*The Englishwoman's Domestic Magazine*, 1 June 1878), he compared the "continuous and very perfect aeration as in the Crystal Palace" with "the comparative inutility" of the air-bubble system as seen (he was speaking of Brighton Aquarium here) in the "turbidness of the water and the fewness as to number and variety ... of the animals and their good health".

In fact, Lloyd's system was sufficiently well engineered to cope with an unexpected rise in the temperature caused by the drying out of the plaster on the new walls and a consequent reduction in the generally cooling influence of dampness (*The Times*, 18 October 1872). This was well recognised by contemporary commentators such as the Orthodox priest Father Nicholas Bjerring writing on the "New York Aquarium and Its Contemporaries":

---

12 One is reminded of Ford Madox Brown's beautiful little portrait entitled *William Michael Rossetti by Lamplight* (i.e. gaslight) held at Wightwick Manor.

> The aquarium is one of the oldest, and is decidedly the best, of the English establishments. It is true that the one at Regent's Park outranks it in age, as do certain others, but owing chiefly to the intelligent management of Mr Lloyd, the water in the tanks of the Crystal Palace Aquarium is more clear and the creatures inhabiting it more vigorous than any we have seen elsewhere.
>
> (*Frank Leske's Sunday Magazine*, 11 September 1877)

By the time the Crystal Palace Aquarium opened in 1871, Lloyd's first masterpiece in the Bois de Boulogne was in ruins, together with the surrounding Jardin d'Acclimatation, as a result of the Prussian bombardment of Paris and the slaughter of the animals that lived there to provide food during the Prussian siege.

Fish came from various places and the company set up "resting depots" for newly acquired stock to acclimatise or to hold in reserve newly acquired stock at Plymouth, where a sea pool was constructed on a joint concession from the Admiralty and the local council, and at Southend. The one at Plymouth was especially valuable. As the *Morning Post* (13 October 1871) put it:

> The Plymouth sea pool furnishes the means so often sighed for by the directors of continental aquaria – which, by the way, are continually increasing in number – of obtaining a regular supply of animals.

In addition, Lloyd had access to the network of collectors and other littoral contacts that he had nurtured since his aquarium warehouse days and he maintained smaller depots and agents at Weymouth, Tenby and Menai, "besides which there are sources whence casual supplies of animals are derived" (*Morning Post*, 13 October 1871). Internationally, the steamers of the Norwegian and Hamburg Line were transporting fish for the Crystal Palace in specially designed holding tanks located directly beneath the ships' bridges (*The Times*, 18 October 1872).

However, the great triumph of the Crystal Palace was to display the first live octopus to have been seen in England and this created a sensation. As John Taylor wrote:

The first living specimen of a living octopus in the Crystal Palace Aquarium had to bear the uninterrupted gaze of lookers-on for weeks. It sat for its portrait in the illustrated papers and had all its points noted down by the newspaper correspondents with the same faithful detail as if they were those of prize cattle at the Agricultural Show.[13]

If anything can be said to have guaranteed the success of this risky new venture, it was the octopus. But why, given that octopuses are found commonly in the coastal waters of the British Isles and France, was it such a sensation to the point where it had its own craze of cephalopodomania? This is especially interesting as, although everyone said (and still says) it was the first octopus to be exhibited in England, a letter from Frank Buckland to *The Times* clearly stated that there was one in the Regent's Park Fish House in 1863 and a cuttlefish (which is often seen as interchangeable with an octopus at this time) had been exhibited in Dublin in 1853.[14] What had changed to transform an animal that before had gone relatively unnoticed into the only thing that every Londoner wanted to see?

The answer lies in Victor Hugo's novel *Les Travailleurs de la Mer*, which appeared in 1866 with an English edition, *The Toilers of the Sea*, published in New York in 1867. The set piece of this novel is an encounter between a man and a huge octopus, which introduced the grasping tentacles and desperate hacking that became the common feature of every underwater adventure and pulp science-fiction cartoon ever since. The whole idea of the tentacular growth of malevolent cities dates from this. It is no coincidence that when, in 1901, Frank Norris wrote of the baleful effect of the railways in southern California, he chose to call his novel *The Octopus*. In 1877, when Frederick Rose wanted to graphically depict the threat of Russia on his "Serio Comic War Map of Europe", he chose a huge and bloated octopus, its tentacles squeezing the life out of Turkey and reaching menacingly towards other

---

13   Taylor, J.E., *The Aquarium: Its Inhabitants, Structure, and Management* (London: Hardwicke and Bogue, 1876), p. 228.
14   *Royal Zoological Society of Ireland, Proceedings of the Society, 1840–1860* (Dublin: The Council of the Society, 1908), p. 18.

countries, both east and west. So, a visit to the Crystal Palace to see the octopus (or Devil Fish as it was commonly known) was not about seeing a small and harmless fish of interesting shape and mysterious habit. Rather, it was about experiencing the harbinger of an underwater world full of monstrous peril and an aggressive enemy of humanity just as dangerous as any lion or tiger.[15]

If we don't count the one that Buckland said he saw in Regent's Park or the Dublin cuttlefish, the first octopus exhibited in Europe appears to have been the one put on show in the aquarium at Boulogne in 1867 exactly when Hugo's novel was being widely read and marvelled at. Henry Lee, who became director the Brighton Aquarium, went to see it and recorded that:

> It was the prominent subject of conversation at the *tables d'hôte* of all the hotels there and almost the first words addressed to a new-comer were "Have you seen the Devil Fish?"[16]

Lloyd outdid the French and obtained ten specimens of two different species. These arrived in the aquarium on 15 November 1871 (*The Standard*, 17 November 1871) and on 20 November Lloyd wrote to *The Times* saying that:

> Their [octopuses] living forms and colours (ever changing) and habits are not at all well known even to naturalists, and now, for the first time in England, they can be minutely and leisurely inspected at Sydenham, where there are about ten specimens ... obtained from Devonshire and North Wales.

---

15   Rose described himself as the "Author of the Octopus Map of Europe". See Ingleby, M., "'Human Language Can Make a Shift': Late Victorian Tentacular Cities and the Genealogy of 'Sprawl'", in W. Parkins (ed.), *Victorian Sustainability in Literature and Culture* (London: Routledge, 2017), pp. 146–64; and Stott, R., "Through a Glass Darkly. Aquarium Colonies and Nineteenth-century Narratives of Marine Monstrosity", *Gothic Studies*, 2 (2018), pp 305–27.

16   Lee, H., *The Octopus* (London: Chapman & Hall, 1875), p. 7.

The craze started immediately. Only five days later, the *Manchester Times* (25 November 1871) reported a soirée of the Manchester Field Naturalists' Club where drawings of an octopus were exhibited (had a member gone to the Crystal Palace to see the new arrivals?) and by 2 December the *Illustrated London News* published a large engraving of "the eight-armed cuttle fish in the Crystal Palace Aquarium". The *Morning Post* (27 December 1871) mentions the octopus as one of the Christmas entertainments at the Crystal Palace and *Reynold's Newspaper* in the same week told its readers of "the wonderful octopus" in what was a facility still sufficiently new for not everyone to have heard of it or really know what it was. In February 1872, *The Gentleman's Journal* recorded that "the octopus ... attracts most attention". The *Pall Mall Gazette* (4 January 1872) recommended the octopus to its readers and *The Ladies' Treasury* (1 February 1872) carried a long essay on "The Octopus" by Eliza Warren Francis.[17]

As with all marine animals, octopuses are delicate and in its review of the aquarium, "The Sea at Sydenham", *The Times* (23 January 1872) mentioned that one of the captives died "about Christmas" and that a sketch had been made of it from its own ink. This was probably the one that Lieutenant-Colonel Stuart Wortley, who claimed to have studied octopuses closely in the wild, mentioned in a letter to *The Times* of 7 December 1871:

> The behaviour of the sickly specimens in the Crystal Palace aquarium (one was lying dead in a corner when I last visited the aquarium) would give very little clue to their motions when in a healthy and wild state.

If we assume that Colonel Wortley saw this octopus on, say, 1 December 1871, which feels near enough to "about Christmas", then it had survived just 16 days in captivity. However, it may have been luckier than the specimen exhibited in York by the Reverend Blake.

---

17   Eliza Warren Francis is another extraordinary Victorian who deserves to be better remembered. See Ridder, J. de and Remoortel, M. van, "Not 'Simply Mrs. Warren': Eliza Warren Francis (1810–1900) and the *Ladies' Treasury*", *Victorian Periodicals Review*, 44 (2011), pp. 307–26.

This was a preserved specimen and as if to illustrate his boldness in capturing this monster of the deep the good vicar recorded that:

> When they had thrown it into a bottle of proof spirit they had to press with their whole weight upon the cork to keep it in.
>
> (*York Herald*, 7 September 1872)

And here is a good example of the different moral status that marine creatures appear to occupy. One cannot imagine an account of such casual and horrible cruelty being applied so callously to a mammal or bird. Over the years I have written about animals there is almost always one incident that I cannot get out of my mind. In the case of this book, it is a vision of a priest and his friends drowning an octopus in alcohol and the distress and pain of the octopus being such that it took several men to keep it submerged.

Lloyd may have felt that he had created a monster, as immediately his specimens went on show, the letters page of *The Times* began to hum with tales of vicious octopuses and people feeling the grasp of voracious tentacles while paddling off Biarritz or Marseilles. On 17 November 1871, Lloyd wrote to *The Times*:

> Allow me to say the octopus is quite as well known to naturalists as a whelk or oyster ... I have never known or seen an example of an octopus attacking a man.

This was a tricky moment as while, on the one hand, Lloyd wanted to spruik the rarity and sensation of his new acquisitions, on the other, he wanted some kind of proportion to prevail in visits to the aquarium. On 24 November he wrote to object to the term "mansucker" that another aquarist, John Lord, had used. "You might as well call horse 'mankicker'" wrote Lloyd, adding, in words that the evangelical Gosse would greatly have approved, that anything which is not actually deformed but presents itself as nature intended it "should be considered beautiful because made for a definite purpose". He returned to this theme on 4 December, pointing out both that octopuses were not inherently aggressive and that the stories of them coming up onto the land cannot be true as, out of the water, they cannot support their own

5 The public aquariums 1

weight. However, it was a difficult balancing act as the very attraction of the animal depended on the projection of danger and aggression, and managing this perception against the cooler claims of science was a process on which a lot of money depended.

The octopus remained a sensation. In September 1872, Boucicault's new farce, *Babil and Bijou*, was staged including a "Cavern of the Octopus" (*Illustrated London News*, 7 September 1872) and at Christmas people could go to see the octopus and then stay in the Crystal Palace for the pantomime, *Jack and Jill*, which included an octopus. Its fame also spread worldwide. The Calcutta paper *Friend of India* (18 July 1872) reported that the Burmese ambassadors to England had been taken to see the octopus. The *Pall Mall Gazette* (30 September 1872) expressed some doubts as the value of aquariums at all and gives us some sense of what the octopus may have looked like:

> The aquariums at the Crystal Palace and at Brighton are only toys after all. Let the fish farmers go to work in earnest. Let them not be satisfied with exhibiting a melancholy octopus but let them render pisciculture a branch of science.

However, in spite of this cautionary note, the octopus carried all before it and it is probably true to say that the initial success of the public aquariums depended on this one "star" animal. As Henry Lee wrote in 1875:

> An aquarium without an octopus is like a plum pudding without plums.[18]

This was a message heeded by William Coup and Henry Reiche when they set up the Great New York Aquarium mentioned in a previous chapter. But the introduction to the *Guide* to that establishment mentions only a preserved octopus and when one looks at the entry on octopuses one discovers that:

---

18 Lee, H., *The Octopus* (London: Chapman & Hall, 1875), p. 7.

We received several of these interesting animals from Bermuda but they all died having been only a few days in the tanks.[19]

The sad fate of a captive octopus en route to an aquarium was summed up by the professional fisherman Henry Haas who was a fish hunter for the New York Aquarium (and had previously served Brighton Aquarium in the same capacity) when he spoke about his most recent consignment of tropical fish:

> I have one octopus in this collection but I feel he will not live. We had a terribly rough passage, and he was so dashed about and against the sides of his can [!] that I am afraid he will die of exhaustion.

(*Indianapolis News*, 1 November 1877)

In his *Guide to the New York Aquarium*, Herman Dorner stresses the difficulties in setting up such a venture and praises the proprietors for sticking at it:

> Then the fish would meet with accidents and die. In one clock beat the beautiful results of a two months' tropical voyage has been rendered useless.[20]

The real accident was, of course, being caught by a fish hunter in the first place.

The establishment of the Crystal Palace Aquarium involved the articulation of finance and capital, legal contracts and rental agreements, cutting-edge technology, manufacturing and engineering, co-ordination of a network of subcontractors, the development of an integrated transport system and, of course, the expertise and experience of William Lloyd himself. As the *Morning Post* (13 October 1871) rightly said, "the aquarium furnishes the first instance in England of

---

19 Dorner, H., *Guide to the New York Aquarium*, (New York: D.I. Carson & Co., 1877), p. iv.
20 Dorner, H., *Guide to the New York Aquarium*, (New York: D.I. Carson & Co., 1877), p. 67.

5 The public aquariums 1

the application, on a somewhat large scale, of engineering to practical zoology". What is more remarkable is that all this was a purely speculative venture for, although it was known that the Fish House had been a sensation and that the European aquariums had attracted significant interest and, at least initially, high numbers of visitors, it was not known that the same kind of success would attend a similar institution in England. What must have given the shareholders the greatest comfort was the fact that the Crystal Palace itself was an established attraction and that they did not rely on the aquarium alone to bring people to Sydenham. This aspect of the aquarium – one attraction among many on a single site – was one of the strengths of the establishment in its early years but, as we shall see, the competing attractions and the gradual diminution of the popularity of the Crystal Palace generally as other recreational options presented themselves in the later nineteenth century also led to its decline. Perhaps ominously, almost as soon as the aquarium opened the *Birmingham Daily Post* (9 September 1871) was pointing to the shortcoming of the overall business model by which an entry fee was charged for the Palace and then extra fees for the separate attractions, such as the aquarium:

> From a long acquaintance with the Palace we know that these extra trifles [except as Lloyd well knew from experience they weren't trifles to the working-class visitor] are generally more or less resented as impositions.

Given that so many things had to be brought together at the same time in order for the project to work, it is not surprising that its opening was somewhat delayed, or rather, delivered in stages. There was a one-off opening on Good Friday (7 April 1871) to enable people at least to see what a large public aquarium looked like. For this event six tanks were filled with fresh water and stocked with fish and only a small charge was made. The partially completed aquarium was then opened more formally on 18 September 1871, as part of a Gala Day which included a tight rope walk by the celebrated Blondin, a concert, a balloon ascent, military bands, canoe races on the lake, archery, cricket, an art exhibition and all the fountains turned on. It must have been quite a day and, as with the original Palace of the Great Exhibition, special excursion trains poured in from all over the country. The *Grantham Journal* (9 September 1871) advertised one such holiday trip. Then on 11 November 1871 the aquarium was finally completed and fully opened and an exclusive soirée featuring speeches by Lloyd's old benefactor Richard Owen and the ubiquitous Frank Buckland – one the most eminent naturalist and the other most popular naturalist of the day. It was a bit like having Elizabeth Blackburn or Jane Goodall share the honours with David Attenborough.

At this point the entry fee was set at 1s but there were more expensive days at 5s and 7s 6d for those who did not fancy mixing with the working class. The aquarium was a great success that more than justified the risk and the investment. In 1869–70 the annual gross attendance at the Palace was an astonishing 2,005,707 (about 70 per cent of the population of London) but the following year, with the added attraction of the aquarium, that had risen to 2,120,822 (which means that an average of 5,810 people were there every day). The revenue from this increase was so significant that for the first seven months of operation the shareholders received a 10 per cent dividend (*The Era*, 16 June 1872) and, as we shall see, it was these large returns on speculative risk that drove a boom in aquarium building throughout the 1870s, often with unhappy outcomes.

Later in the year the *Morning Post* (11 December 1872) reported another shareholders' meeting which was told:

> The aquarium continues to justify the favourable expectations expressed concerning it in former reports and is undoubtedly a source of considerable attraction to the public.

What this attraction consisted of was well expressed in an article published in the *Birmingham Daily Post* (9 September 1871) even before the aquarium was fully opened:

> What hot and dusty mortal, condemned to spend these August afternoons in the heat of London, wouldn't feel it a treat to realize what it must be to be a lobster, sitting in the mouth of his own "cool grot" right down amongst the rocks, and in the still deep shades of the briny? This is a treat which may, to a very great

extent, be now enjoyed at the Crystal Palace, where, within the past few days, the great marine aquarium has been opened.

And certainly it was popular; numerous newspapers reported on the crush of visitors and the *Daily News* (4 April 1874) noted that it was very difficult for children to see the fish because of the great crowds of adults standing in front of the tanks.

Among the many accounts of visits and excursions, one stands out and that was the visit of the Worcester workhouse boys (it appears that the girls went too but they are barely mentioned), which was laid on as a treat by the guardians and various charitable individuals and was reported on by the *Worcester Journal* (5 October 1872). The children were taken by a very early train to London at 5s for adults and half a crown (2s 6d) for the children and thence by local train to the Crystal Palace at 1s 3d for adults and 7½d for the children. This included admission to the Crystal Palace and they were then given a discount on admission to the aquarium. This was all paid for by the guardians of the workhouse and public subscription. The children were each given 6d, which, it is recorded, they mainly spent on the little toys (pretend watches were especially popular) from the stalls in the Palace. A packet of sweets, an apple pie, some ginger beer, fruit, plum cake and ham and beef sandwiches were all provided by local shops. The boys were enjoined especially to look at the scientific and technical exhibitions to inspire them for what they might achieve in life (the girls just had to hope for the best). After a long day they were taken back to Kensington Station for a bun and a glass of milk, whence they boarded the return train to Worcester arriving in the early hours of the next morning. The children's responses to the aquarium aren't recorded but for most of them the experience must have been of a novelty so strange (remember that most of them, coming from landlocked Worcester, would never have seen the sea) that it was almost impossible to understand. And this incident does offer a little variation to the general narrative of the unrelenting brutality of the lot of the workhouse child in Victorian England. It also contrasts with the account of a visit to the aquarium for an upper- or middle-class child described as "Trots with the Editor to the Crystal Palace Aquarium" in the illustrated children's magazine *Little Folks* (no date but issued in 1872).

Lloyd wrote a description of the Crystal Palace Aquarium in *Nature* (4, 1871) and another in the *Athenaeum* (1 April 1871) and in these he not only described the collections but also stressed the engineering of the pumping and circulation systems, which were the key to the success of the establishment not only in keeping the tanks healthy but also in maintaining large numbers of a variety of species in each tank. Thus managed, the sea-water tanks remained healthy but when the freshwater tanks were added they proved more problematic. *The Times* (18 October 1872) noted that these were stocked "only when local filtration can purify the contaminated London water". Interestingly, it was possible to purify the water, as the filtration of the now largely forgotten Floating Swimming Bath Company's establishment moored on the Thames adjacent to Charing Cross between 1875 and 1885 seems to show. This was a floating Crystal Palace, containing a swimming pool 135 feet long and 25 feet wide. It was fed from the Thames by forcing water – which had been allowed to settle in tanks through a layer of sacks filled with gravel and sand – at the rate of 500 gallons per hour and of sufficient clarity to enable people to swim without contracting a serious illness (*Illustrated London News*, 17 July 1875). In winter, the bath became the Glaciarium and was frozen for a skating rink (*The Times*, 20 December 1876). It was discovered that, in London, the Thames had a high proportion of sea water and, at high tide, was almost all sea water, so it may be that this chemistry was a contributory factor to the problems of locally filled freshwater tanks at Sydenham.

Several years later Lloyd contributed a short piece called "Notes on the Structure of Aquaria" to *The Zoologist* (11, 1876) where he explained that the key to his method was control of the water temperature:

> Our great success at the Palace depends very much on our temperature being nearly that of the English ocean in all seasons and it is this, conjoined with complete and constant aeration by our machinery, that enables us to keep in a comparatively small space so many animals, and many of them of kinds which, once we received them uninjured, are maintained nowhere else under the same inland conditions.

These included herring – which are apparently notoriously difficult to maintain in captivity – but alas the congenial water temperature also suited a voracious eel that ate them together with the sepia cuttlefish.

By 1875 the attractions of the aquarium were starting to decline. Visitor numbers were falling off and the *Morning Post* (21 December 1875) reported that the aquarium company was attempting to shore up its finances by investing in the Crystal Palace itself. The next year the *Edinburgh Evening News* (9 June 1876) reported that the aquarium had "fallen on evil times", there were schisms in the board and the shareholders had received no dividend. This was mainly due to a decline in visitors to the Crystal Palace itself as rival attractions in London, especially the Alexandra Palace and the Royal Aquarium newly opened in Westminster (this is considered in more detail in the next chapter). There were still good days. Massive crowds were reported on Whit Monday of that year (*The Era*, 11 June 1876) and the next year *The Times* reported that, although there were disputes on the board over how much to spend on repairs to the fabric, the shareholders still received a dividend of 10 per cent.

But the end was in sight for this pioneering aquarium. In 1878 Lloyd left, as the company could no longer afford to pay his £400 per annum salary, and the company found it hard to replace him with someone who would be competent to manage such a large and complex facility. The various editions of *The Crystal Palace: Guide to the Palace and the Park* (also entitled *General Guide to the Crystal Palace and Its Gardens and Park*) continue to refer to the aquarium into the late 1880s but the edition of 1878 tells us that you still need to buy Lloyd's *Guide* separately (by this time, it must have been quite inaccurate) and also that another small aquarium had opened in the stalactite cave that was part of the dinosaur exhibition and for which you had to pay an extra admission charge. From this point on we see a sad decline punctuated by attempts to improve things. The report of the aquarium company meeting in *The Standard* (1 August 1891) refers to an outlay of £498 to put the aquarium in good order:

> It had been a matter of consideration with the board to know what to do with the aquarium, whether to allow it to get into a bad state or to repair it ... Rightly or wrongly they had placed the aquarium

in thorough condition, and the public appreciated what they had done because they largely patronized it.

The *Morning Post* (14 August 1891) reporting on the same meeting added "they had now got a good man to manage it". But good man or not, the aquarium slips from view in 1892 and by 1894 the aquarium terrace is mentioned only as a concert venue. It appears that at about this time the long-running monkey collection that had been a feature of the Palace almost since its move to Sydenham was relocated into the aquarium tanks and where herring and octopus once swam in sea water of an optimum purity and temperature monkeys now gambolled.

The aquarium found a new lease of life at the very end of the century when it was decided to set up a School of Experimental Pisciculture at the Crystal Palace (which itself was an ailing facility by this time). The *Fishing Gazette* (14 October 1899) reported that:

> The aquaria are in working order, and in conjunction with the excessive reserve tanks behind, now form a capital home for the complete collection of British fresh-water fishes it is intended to place in them.

Later that year the *Sporting Times* (16 December 1899) was reporting that:

> The authorities [i.e., the management of the Crystal Palace] place at the use of the school the whole of their magnificent aquarium, tanks, ponds, fountains, basins and lakes etc.

Here is perhaps the greatest evidence of the decline of this once astonishing popularly institution: the fountains that were a key attraction of its galas and the lakes that once hosted canoe races had now been converted into fishponds. The school did, however, enjoy some success and *The Times* (14 March 1901) noted that fish culture was actually now taking place at the Crystal Palace.

The Brighton Aquarium was founded just after the one at the Crystal Palace and on a similar basis. A joint stock company was set up and this raised and invested the very large sum of £100,000. In the early

5 The public aquariums 1

years, it paid a generous dividend. The seaside location meant that the aquarium did not have to transport water over long distances but the pollution of the offshore environment coupled with the decision to keep the tanks pure by using aeration rather than continuous circulation meant that this brought its own problems. The first manager was the marine scientist and associate of Frank Buckland, John Keast Lord, but, not long after the aquarium was opened, he suffered a stroke and subsequently died. He was replaced by the equally experienced microscopist Henry Lee who brought in, probably on Frank Buckland's recommendation, William Saville-Kent from the British Museum as "resident naturalist". It was Lee who was mainly responsible for the initial stocking of the tanks, although it appears that, once the aquarium was open, he was an infrequent visitor even though he was nominally in charge. As at the Crystal Palace, the aquarium took some time to get going, leading, as the *Morning Post* (5 February 1872) reported, to shareholders' meetings in which "discussions ... seemed likely at times to take a rather recriminatory turn". It had what we would now call a "soft opening" on 1 April 1872. The event was graced by the presence of Prince Arthur who must have noted the date as significant for the inauguration of an aquarium with no fish. A month earlier the *Daily News* (14 March 1872) had presciently observed that:

> It will not, however, be easy to people the tanks with fish by that time [i.e., the planned opening], and to all appearances, if Easter visitors are lucky enough to see water in the tanks, they must fain take the fish for granted.

The official opening was then set for 27 July 1872, but this was postponed until 12 August when the company realised it could take advantage of the huge number of delegates to the Congress of the British Association of Science which was taking place in Brighton that week.

The aquarium actually opened on 10 August 1872 and, like the Crystal Palace, was an enormous establishment compared with anything that had been seen in England before. It cost £50,000 just to build with a total project cost of £133,000 as it was necessary to construct not only the aquarium building but also a new sea wall and promenade. The Brighton Aquarium was, like the one at the Crystal Palace, largely underground. The architect Eugenius Birch built a catacomb of large arches that blended the Gothic with the then-fashionable Pompeian style. The building was, innovatively for the time, provided with ramps so that people in invalid and bath chairs could access it and had 41 tanks, the smallest of which was 11 by 10 feet and held 4,000 gallons of water and the largest of which was an enormous 130 by 30 feet and held 110,000 gallons. There was no need for the huge reservoirs that kept Lloyd's continuous circulation system in motion but, instead, each tank had its own air pump. There were various other facilities and these were augmented over the years but, at first, they included a study room with books, periodicals and manuscripts. In 1874 a roof terrace, which was part of the original plan, was completed together with a clock tower and ticket offices, and this terrace was significantly extended in 1876 to accommodate a roller-skating rink (another Victorian craze), a smoking room, a billiards room café and a conservatory for music. Accounts of days out in Brighton later in the century often mention this terrace as a place of popular resort. Emma Southwick's piece "The Aquarium at Brighton" in the American magazine *St. Nicholas* (1 August 1879) captures the spirit very well:

> Many poor families in London save up their pennies till they can buy excursion tickets to Brighton and down they come by thousands, to spend the day on the beaches, to see the aquarium and sit here on the flat roof and smell the sweet breath of the salt sea [seasoned with a top note of raw sewage], watch the gulls wheel about overhead, and see and hear all the charming sights of this charming place.

And a South African visitor writing in the Cape Town magazine *Lantern* (20 December 1879) said that the aquarium was "fairly crowded with visitors at all hours of the day" and characterised the entrance hall as "a sort of club [where] you may read the Fashionable Visitors List". It appears that this list was self-selecting, so, although it might include a European monarch or a member of the British royal

5 The public aquariums 1

family, it might equally include a Pooterish alderman and his family down on a visit from Holloway.

Even at the opening, the tanks were still understocked. The *Birmingham Daily Post* (12 August 1872) reported on the opening and told its readers that:

> There were but few large fish – a small shark had been bought from local fishermen the night before but had promptly died.

Frank Buckland was one of the speakers at the opening and amused his hearers by fishing a small crocodile out of a box he was carrying. The *Daily News* (10 August 1872) pointed out that the aeration system would make it difficult to maintain the tanks and that on one of them the glass had cracked due to the contraction of its iron frame. Two days later the same newspaper reported that the mayor – who had officially declared the aquarium open – was heckled at the ceremony owing to what was perceived as the obstructive attitude of the local authority during the planning and building phases. But there was obviously nothing personal as he subsequently joined the board and became a shareholder.

But the town and the company now had an aquarium and it hosted its first VIP visitor on 19 August. This was the now-exiled Emperor Napoleon III who joined a throng of 5,200 other visitors that day (*Illustrated London News*, 7 September 1872). A week later there were 6,000 visitors at an entry fee of 6d (to give a revenue of £150 for the day from entry fees plus whatever was made from the sale of refreshments and merchandise, so, although the capital expenditure had far exceeded Lloyd's recommendation for sustainability, the revenue was at this point much more than the minimum he considered viable for an aquarium run on commercial lines). The *Illustrated London News* (7 September 1872) reported that some progress had been made in stocking the tanks and gave extended descriptions of a number of the fish. The aquarium also acquired the all-but-indispensable octopus which was retrieved, prosaically for a prodigious monster of the deep, from a lobster pot just off the beach at Eastbourne.

But from the earliest days all was not well. Saville-Kent had a public falling-out with Lee over the latter's publication of what Saville-Kent considered to be his own research on the sexual behaviour of octopuses. On the opening day, the *Daily News* had observed:

> Yet with all the accommodation provided for them in the Aquarium, it may be doubted whether the best conditioned fish there are wholly happy. It is still no better than a splendid prison.

A couple of months later *The Times* (18 October 1872) pointed to the poor aeration system and the water in the tanks cloudy with sewage, blaming:

> the vicinity of the sea [which] has caused the designers to be careless of the value of sea water, when once acquired and properly purified so as to be serviceable for aquaria.

Lloyd had been a consultant on the design but his advice on circulation was not taken. Father Bjerring (*Frank Leake's Sunday Magazine*, 11 September 1877) balanced his praise for the Crystal Palace with less flattering remarks about Brighton:

> If the management of the Crystal Palace Aquarium is such as to merit our approval, that at Brighton does not appear so commendable. In truth, the Brighton Aquarium is unattractive in respect to the appearance of the water in its tanks. This is opaque and dense, a serious defect, due to the stubborn adherence to the ancient system of aeration by air currents only [i.e., not by continuous circulation] and until the advice of experts is followed, the famous Brighton Aquarium will not take the rank to which it is entitled.

But these critical comments were nothing compared with Saville-Kent's blast in *Nature*. He also criticised the inadequate aeration system (note that he thought that aeration was a viable option for large-scale aquariums and was here criticising the particular installation at Brighton, not the system itself) and pointed out:

It is impossible, for instance, to keep in health at the Brighton Aquarium, the number of fish in comparison to the size of any given tank as will be found in the aquarium at the Crystal Palace.[21]

He gave the examples of the very large tank No. 6 which contained 110,000 gallons of water yet contained only three dogfish, a ray and some turtles and of tank No. 11 which contained 9,000 gallons yet only contained two mackerel. Saville-Kent was also scornful of the potential of the Brighton Aquarium as a scientific institute, adding that there was no contradiction between an aquarium with a primarily educative mission and a commercially successful business:

> The Isle of Wight, and the Devonshire coast, especially Torquay, are locations offering far greater advantages than Brighton, as zoological stations for the acquisition of specimens, and now that the financial success of large aquaria under judicious management is well established, the temptations these places offer to an enterprising company cannot be long resisted.[22]

However, these squabbles between marine scientists and the internal management conflict of the aquarium did not affect the popularity of the institution as a place to visit. A review of Saville-Kent's *Official Guide-book to the Brighton Aquarium* in *Nature* summed up the situation trenchantly, "it shows that Science will not be entirely neglected in the endeavour to attain material prosperity".[23] As *The Spectator* (8 January 1875) put it:

> Brighton Aquarium is not a great success, considered zoologically, though, as a show, it is entirely successful.

The aquarium had the great advantage of being in a town that was already both a fashionable and a popular resort, especially for

---

21   Saville-Kent, W., "The Brighton Aquarium", *Nature*, 8 (1873), pp. 531–3.
22   Saville-Kent, W., "The Brighton Aquarium", *Nature*, 8 (1873), p. 532.
23   Anon., "Our Book Shelf", *Nature*, 8 (1873), p. 160.

Londoners, and so it could trade off the enormous numbers who would be in Brighton anyway and were always looking for something new to pass the time and, given the uncertainties of the English weather even in summer, an undercover entertainment was especially welcome. As Conlin has noted, using Brighton Aquarium as an example:

> By the 1870's the divide between "useful knowledge" and entertainment had entirely disappeared and not just in London.[24]

The initial popularity may be gauged by the first six months of revenue. This was reported by *The Standard* (27 February 1873) as amounting to nearly £9,000 yielding a profit of £2,769 6s 6d, which was ample to support a 10 per cent dividend for the shareholders.

With the aquarium now fully opened, there were band performances three times a day and promenade concerts on Saturdays. More VIPs, the King and Queen of the Belgians and the Shah of Persia visited in 1873 and *The Times* (7 June 1873) reported that 125,000 and 130,000 people had visited on two consecutive days of a holiday weekend and takings were between £600 and £800 per week (*Sheffield Daily Telegraph*, 20 September 1873). This commercial success prompted the aquarium company to seek permission to raise a further £50,000 in capital and this was granted. However, it was decided to raise only £30,000 through the issue of shares at 10s each and yielding five per cent interest. The decision was justified by the continuing numbers of visitors. *The Times* (1 January 1874) reported that 4,160 people had poured in on the evening of New Year's Eve 1873 and by February the *Morning Post* (27 February 1874) reported 10 per cent dividend payments for each half-year of 1873 with the same guaranteed for each half-year of 1874. The new investment enabled the extension of the terrace mentioned above, the addition of a fernery and an innovative message board where news telegrams from different parts of the UK were posted throughout the day, enabling people to see what was going on around the country more or less in real time. In addition, the railway companies were now running regular aquarium excursion trains from London with the cost of entrance included in the

---

24 Conlin, J., *Evolution and the Victorians* (London: A.C. Black, 2014), p. 188.

ticket price or, in other cases, offering an arrangement by which cheap admission was granted on production of the rail ticket.

However, the popularity of the aquarium was also a cause of controversy. The building had always been open on Sunday. It justified this by its educational mission and by ensuring that the band only played sacred music and hymns. This caught the eye of the Sabbatarian movement and in 1875 a case was brought against the aquarium under the provisions of the *Lord's Day Observation Act* of 1780. This had originally been passed to try to prevent subversive gatherings during the Gordon Riots but was still in force and was now used to try to restrict social entertainments on Sundays. The long-running sequence of cases concerning the aquarium are complex and not of interest in themselves except insofar as they expose another aspect of the campaign to control working-class leisure. As the *Sheffield Daily Telegraph* (3 May 1875) trenchantly put it:

> It is essentially a poor man's question: to the working man Sunday is everything.

However, the cases are of interest as they also show how the aquarium tried to present itself and some of the thinking that had gone into the framing of its mission and purpose.

The aquarium company lost the first case and was made liable for a £200 fine on every Sunday that it opened. *The Times* (28 April 1875) reported that the judges were unhappy with their own verdict:

> Both judges found for Terry [the plaintiff] but one said he wished he could find the other way; and the other said that although the law was clear, the kind of wholesome and instructive entertainment at the aquarium should be encouraged.

This shows that the defence posted by the aquarium company that, although the aquarium was a place of entertainment which charged an entry fee (and thus illegal under the Act), what it offered was wholly consistent with the respectable conduct of a Sunday in a Christian nation. In other words, although the aquarium had, in practice, become as much a place of general entertainment as an educative display of marine creatures, it still placed its scientific mission at the centre of its self-image and, as we shall see, this is characteristic of later aquariums at least one of which appears to have had no fish.

One anxiety that was raised over this case was that it might set off a domino effect that would lead to the Sunday closure of the zoo, the Botanical Gardens, the Horticultural Gardens and the band concerts in Battersea Park. And, ironically, although all these eminently respectable and edifying places of resort were now under threat, pubs could open quite legally. *Punch* (15 May 1875) published a piece entitled "Fanatics vs. Fishes" and a cartoon showing two obviously drunk men outside a pub on a Sunday with one saying to the other "Thish is my aquarium". The aquarium company proposed that it could open on a Sunday simply to show the collection of fish but not hold the concerts of sacred music. However, this was not acceptable as the sticking point in law was paid admission. In the end, and after further cases in 1876 and 1877, the home secretary himself intervened with a compromise that suited nobody but did offer a way out. He decided that he would not change the law but would remit any penalty the aquarium company might face for flouting it. And in 1879 this was confirmed by a royal warrant.

What was at stake for the aquarium was not just the claim that its primary mission was to educate the public and advance the cause of marine science; there was also a great deal of revenue in jeopardy should Sunday openings have to cease. In 1874, for example, receipts for Sunday opening came to £2,815 (*Leeds Mercury*, 14 February 1874). So, although the Sunday openings were now tolerated, the company would clearly have preferred to de-risk this revenue stream entirely by a change in the law.

Brighton has always had a reputation for a raffish loucheness and the aquarium sometimes attracted unwanted publicity because of this. On 22 November 1875, the *York Herald* reported that:

> Mr Bennett, secretary of Leeds Town Hall told his wife and family that he was going to be interviewed for the secretaryship of Brighton Aquarium but disappeared.

It turned out that there was no such job on offer and Mr Bennett had serious gambling debts. He was last heard of on a boat to the United

States accompanied by a "young unmarried female". Very Brighton. Aquariums, like zoos and menageries, could also be sites of crime and disorder. The *Morning Post* (2 October 1877) told its startled readers that not only had four baby turtles been stolen from their tank but that an enterprising ruffian had put an electric eel in with the alligator.[25]

By the end of the 1870s the aquarium was starting to lose its attractiveness in spite of the addition of a manatee, sea lions, porpoises and giant turtles to the collection and the appointment of the very distinguished ichthyologist Francis Francis as resident naturalist. The *Leeds Mercury* (14 February 1879) reported that, although revenue had increased because of the many new attractions that had been added to the aquarium entertainment complex, the cost of building and maintaining these meant that, relatively speaking, the company was no better off and although a dividend of five per cent was still payable this did not compare with the high returns of earlier years. In 1879, Monday excursion trains were temporarily suspended and the 1880 shareholders' meeting received the grim news that profits were down £1,140 on the year. This was attributed to the worst summer weather in 20 years but it meant that no dividend was payable to the original investors. Francis Francis, who had £3 invested, suggested that the company could turn things around by allowing advertising inside the aquarium and allowing volunteer soldiers in uniform admission at half price over Easter when their units were training in the countryside around Brighton (*Morning Post*, 13 February 1880).

However, the company could not turn things around and only a week later *Funny Folks* (20 February 1880) referred to "the dismal

---

25 My own research on Victorian zoos and menageries shows that they were the site of an unusual number of crimes and disorderly conduct. My paper "'The Menagerie as Crime Scene" was delivered at the University of Lincoln's *New Contexts for Sweeney Todd* conference in 2008 but has never been written up for publication. An analysis of hundreds of newspapers printed between 1837 and 1901 revealed that reports of crimes committed at menageries were more frequent that reports of crimes committed in any other public place. Whether this reflects the specific opportunities for crime afforded by an animal display (certainly they offered a unique opportunity to attack an exotic animal) or the greater sensation afforded by the opportunity to combine criminality and wild beasts in one piece of journalism is a matter for conjecture.

situation at the aquarium". In 1882 the board received a petition to wind up the company (*Morning Post*, 22 April 1882). This was lodged by one of the directors who wanted payment of a £200 loan and a £3,100 bond. This petition stood for three weeks but did not succeed. However, later that year, an extraordinary meeting of shareholders precipitated a complete spill of the board with the comment that:

> if properly managed there was no reason why the shareholders should not have a fair return for their capital. (*The Times*, 9 May 1882)

But whatever the shortcomings of the management – and it appears that the main issue was lack of continuity – the fact was that the days of establishments like the Brighton Aquarium were coming to an end and, although there would still be plenty of visitors, the numbers required to sustain the business and deliver the returns on capital invested that were necessary for the shareholders were simply no longer there. As with the Crystal Palace, rival attractions and changes in leisure patterns were taking people away from large aquariums, which now seemed quite old fashioned even though, in Brighton's case, it had only been built ten years earlier.

In 1882 the *Sunday Times* carried advertisements for Boxing Day excursions to the aquarium from Brighton at only 1s 6d return with a half-price admission and these were repeated in 1883 and 1884, although by 1884 the price had gone up sharply and a basic ticket now cost 12s 6d with an extra half a guinea in the Pullman car and the same for first class with an aquarium ticket thrown in. It is doubtful that prices like these would have attracted the mass public that the aquarium urgently needed. In October 1885 the company applied to the town council for a theatrical licence in order to diversify their offer (*Standard*, 2 October 1885) but this didn't help and the dividend for 1886 was paid at only two per cent (*Standard*, 17 February 1886). The company tried pantomimes and rose shows but the business continued in gradual decline.

In 1893 the shareholders' meeting was asked to consider a complete financial restructure and registration under the 1882 *Companies Act* and an unsuccessful attempt was made by the board to raise a further capital sum from the long-suffering shareholders in order to enlarge the auditorium so that it could house a bigger paying audience (*The Era*, 25 February 1893). The failure of this attempt led two months later to a proposal to liquidate the company and restructure the finances of a new enterprise to include a further capital investment of £3,000 for the work to the auditorium. The report of this meeting reveals some interesting attitudes among the shareholders. Some agreed that the way forward was to diversify and to develop the general entertainment business. Others argued that there was no point in trying to compete with the music halls and that a better strategy would be to revert to a main business based on the aquarium's fish but to halve the entry fee (*The Era*, 22 April 1893). This is the first suggestion that the price of entry was a contributory factor to the decline in visitors and may show that, in a market where the aquarium was increasingly having to hold its own against other attractions that were not there in 1872, consumers had become more price sensitive.

The company went into voluntary liquidation in 1896 but was still marginally profitable and had one last hope, which was to be purchased. On 9 December 1898, the *Daily News* reported a shareholders' meeting that was considering the offer of a financial consortium headed by a Mr Morton. This consortium would offer £42,000 for the buildings, pay off any liabilities, refund in full the investments of debenture holders and pay off shares at the rate of 1s 5d per share. The company then intended to spend between £30,000 and £35,000 developing a top-class music hall, a theatre and winter gardens. The meeting unanimously took the offer and *The Times* (20 February 1899) reported that a new company, the Brighton Aquarium and Winter Garden Ltd, was seeking subscriptions to raise capital of £165,000 and offering a five per cent return on debentures. Sadly for the shareholders, this venture fell through so, in 1900, the aquarium was sold to the town council for the meagre sum of £30,000. At this point the company would have made a loss of £1,100 and was only kept in the black by the £1,750 that Mr Morton's consortium had lodged as a deposit and that was forfeit when their project fell through (*Nottingham Evening Post*, 12 April 1900). On 31 December 1901, the company was finally wound up and after meeting all liabilities it was left with £5,000, which seems a tiny sum when compared with the massive capital investments of the preceding 28 years (*Evening News*, 31 December 1901).

The Crystal Palace and Brighton aquariums were very different in context and conception but their life cycles as businesses were remarkably similar. What we see in both is the decline of a particular kind of leisure pursuit as competing forms (for example, professional football) grew. I doubt if the mass public for aquariums was ever as interested in the educational and scientific missions of these establishments as the boards and managers hoped. But people did like to see amusing and interesting fish. However, ultimately, they only wanted to see them once or twice and once the tradition in a family of visiting the aquarium on a Bank Holiday, say, faded away there was no one new to replace them. What we also see in both aquariums is over-capitalisation against over-optimistic revenue projections and the problems inherent in failing to maintain continuity of a high-quality management. Perhaps these are always features of the popular capitalism that was made possible by the financial liberalisation of the capital market in mid-Victorian England. Once Lloyd was dead (in 1880) there really was nobody else in England capable of running a large-scale aquarium. Saville-Kent, who might have been given the right funding, had by this time gone to seek his fortune in Australia where we will meet him again in a later chapter. And both aquariums relied on a specific context – the Crystal Palace, the town of Brighton – to deliver their visitors and so were vulnerable to shifts in the popularity of these larger entities as nobody seemed to believe that an aquarium was, in itself, capable of attracting a mass public.

What is missing in all this are, of course, the fish. Throughout the twists and turns of company history they were always there, swimming in their clear circulated tanks or their muddy aerated ones. They are part of the story of the two great public aquariums but as commodities within the systems of exchange which kept the institutions going primarily as financial entities. Ironically, when we speak even of the biggest establishments ever built to display them, we find ourselves continuing also to speak of the strange case of the missing fish.

6
The public aquariums 2

In the previous chapter we saw how two large-scale public aquariums were set up in the 1870s and how they enjoyed much initial success but subsequently experienced financial and other problems. However, that initial success and, especially, the substantial dividends paid to the shareholders of the Brighton Aquarium encouraged investment in public aquarium projects around the country. This chapter will look at these in general and at some with a more granular approach so that the detailed operations and the fine textures of a public aquarium project may be better understood.

This chapter is as much about missing fish as about fish on show. What underlies this aspect of the Victorian encounter with fish is their position – and that of other marine plants and animals – in a chain of capital exchange and as terms in a specific business model. The aquariums of the 1870s varied in quality and in the style of their aquatic collections. However, they almost all have in common – even those with an explicitly scientific and educational mission – the appearance that the fish were secondary to the financial structure of the businesses for which they were purchased. Frequently the outcome for the fish was not good at all, as tanks were allowed to fester and experts who might have been able to keep them healthy were not employed or simply not available.

The 1870s saw an astonishing boom in aquarium projects with companies, sometimes in partnership with local government, raising funds to build aquariums in Aston (Birmingham) (1876), Bolton (1874), Blackpool (1876), Brighton (1874), Derby (1876), Douglas (1876), Edinburgh (1875), Glasgow (1878), Great Yarmouth (1875), Hastings (1875), New Brighton (1876), Llandudno (1877), Liverpool (1875), London (1875), Manchester (1872), Margate (1876), Morecambe (1875), Ramsgate (1877), Rhyl (1875), Rothesay (1875), Scarborough (1876), Southport (1874), Southsea (1875), Sutton Coldfield (1879), Tynemouth (1878) and other places. As Reiss points out:

> the geographic specific of Great Britain as an island with a big ratio of coastline to area and the resulting short distances between coast and inland cities simplified their [marine aquariums] set up and maintenance.[1]

I am not sure that "simplified" is the right word. Perhaps gave "false encouragement that set up and maintenance would be simple" is a more accurate formulation. Of the 25 locations given above, 17 are coastal and of the remaining eight, three (London, Edinburgh and Glasgow) were significant inland ports.

Of these aquariums some, like Manchester, Southport and Great Yarmouth, were conceived of on a very large scale that required significant capital investment while others, such as the modest project at Derby, involved the local council planning to include an aquarium in the new Derby Free Public Library and Museum (*Illustrated London News*, 4 April 1876). At Liverpool it was proposed to set up a small aquarium in the Picton Library. This was to be funded and developed by the city council at the relatively modest cost of £2,800 and the *Illustrated London News* (6 February 1875) reported that an Alderman Bennet had personally offered £3,000 for this purpose. This appears not to have happened as *The Times* (4 October 1879), reporting on Lord Derby's

---

1   Reiss, C., "Gateway, Instrument, Environment: The Aquarium as a Hybrid between Animal Fancying and Experimental Zoology", *International Journal of History and Ethics of Natural Sciences, Technology and Medicine*, 20 (2012), pp. 309–36, p. 314.

opening of the Picton Reading Room, observed that it was originally intended as an aquarium. At Leicester, as early as 1856, the curator of the city museum announced that "a marine tank has been established and a fresh water tank" (*Leicester Mercury*, 29 March 1856). There was eventually more than one aquarium in Blackpool and some of the projects listed above did not come to fruition. Early notions of an aquarium at Kew and one at Bath (which was reported in the *Morning Post* as being planned in June 1855) also came to nothing.[2] In addition, these various aquarium projects operated on the basis of differing business models. Some, like Manchester, had science and education as their primary motive; others, like Southport, were located alongside a variety of entertainment facilities along the lines of Brighton.

There were some similarities between the zoo boom of the 1850s and the aquarium boom of the 1870s. Institutions founded during both booms had a range of missions from the purely scientific and educational to the unabashed pursuit of entertainment. The success or failure of both kinds of institutions appears to have depended on similar factors. One of these was the ability of the proprietors to extend the range and variety of their holdings so that there was always a new or unusual animal to see. At Regent's Park, David Mitchell had understood this well when he instituted the "star animals" system and from 1850 onwards ensured that people would return time and again to see the current novelty. Where aquariums were concerned, we see that Brighton and Crystal Palace were constantly adding to their stock, with the Crystal Palace octopus being the first undoubted "star" animal. But not all zoos or aquariums had the cash, the expertise or the networks

---

2   Blackpool's first aquarium was opened in May 1872 by the Raikes Hall Park, Garden and Aquarium Company and was part of a large entertainment complex. However, the aquarium and gradually the whole of Raikes Hall Park was outcompeted first by Dr Cocker's Aquarium, which was in the town and was eventually at the bottom of the celebrated Blackpool Tower, and then the massive Winter Gardens development. Cocker's Aquarium was an early example of grotto architecture and was modelled on the limestone caves of Derbyshire. It survived until 2010. See Brodie, A. and Whitfield, M., *Blackpool's Seaside Heritage* (Swindon: Historic England, 2015) and Pearson, L., *The People's Palaces: The Story of the Seaside Pleasure Buildings of 1870–1914* (Northampton: Quotes Ltd., 1991).

to do this and this constituted an often fatal weakness in their business model as people simply wouldn't pay to see the same collections more than once. However, whereas a constant supply of exotic animals was pouring into the country and a variety of animals was always available or it was possible to commission an expedition to capture a particular animal (such as a hippopotamus), it was not so easy to acquire a specific fish and it is notable that, where such an expedition was mounted, it was to acquire Californian sea lions which were, like African animals, visible to the hunters in a way that fish were not.

Success also depended on the availability of other attractions. These could be in-house, such as the extraordinary range of entertainments available at Belle Vue or the great spectacles and concerts staged at the Surrey Gardens, or they could be available within the vicinity. Thus, the success of Crystal Palace and Brighton very much hinged on the fact that there were plenty of other things to see or do elsewhere in the Palace or elsewhere in the town. Sometimes another of the aquarium's facilities superseded the aquarium as the main business. This happened at Hastings where the opportunities afforded by the public baths, which were originally just one part of the whole aquarium project, were such that the aquarium was abandoned in favour of the baths. At the public meeting called to debate the project, it was determined that while baths were necessary for the town an aquarium was not. The *Hastings and St. Leonards Observer* (9 January 1875) recorded the comment of Mr Coleman who provided by far the clearest sighted and sanest view of the aquarium boom that I have found anywhere except perhaps in William Lloyd's comments on capitalisation quoted in a previous chapter:

> There were aquariums everywhere and people were getting tired of them; they were all the same. Brighton Aquarium was successful but it was principally through its concert room. Then there was great expense attached to the aquarium; it required much money to keep it up, and keep alive the specimens and renew those that died.

This was in early 1875 when many projects had yet to get under way – it's a pity that every town didn't have a Mr Coleman. And you can almost hear the sighs of relief at the Annual General Meeting of shareholders reported in the *Hastings and St. Leonards Observer* (22 March 1879) when it was pointed out that not proceeding with the aquarium had saved £15,000 and that baths were good business too. In 1879 a five per cent dividend was paid and by 1885 when business was slacker (as more people got their own bathrooms) shareholders were still getting two per cent. The baths were probably more popular with the citizens too – *The Times* (6 July 1875) reported rioting in Southsea over the aquarium and skating rink company's erection of hoardings that blocked off access to the beach. Finally, the success of a zoo or an aquarium crucially depended on its admission price and, in particular, the scale of its prices at the lower end where the working class – who constituted by far the most numerous potential visitors – had to be able to afford to get in. We have seen how even a respectable working man like William Lloyd could only access the Fish House after a personal appeal to the kindness of Richard Owen and from the 1850s onwards there is constant debate not only about admissions prices to zoos and aquariums but also about opening hours that would be convenient to the limited leisure time of the working class, and, in particular, to enable working-class families to enjoy a day out (this included, for example, the provision of women's lavatories – a thing against which the British Museum held out).

However, there were also many crucial differences between zoos and aquariums and these differences also impacted on success or failure. The first was that zoos competed not so much with each other as with the extensive travelling menageries which toured the country, criss-crossing it with huge caravans of horse-drawn wagons containing everything from elephants to pelicans. At its height the biggest of these, Wombwell's, was travelling with nearly two thousand animals. These set up in country towns, often alongside established seasonal fairs (such as the Nottingham Goose Fair), and they were a major feature of the Victorian landscape.[3] Aquariums had no such direct competition and instead had to persuade people to part with a hard-earned shilling to

---

3   See Simons, J., *The Tiger That Swallowed the Boy: Exotic Animals in Victorian England* (Faringdon: Libri Publishing, 2012), for an extensive treatment of travelling menageries.

look at fish rather than buy a book or go to the music hall. Zoos could be in any place in the country and were, to some extent, independent of their immediate environment provided they had good transport links. Most aquariums were in coastal towns and benefited from this location as they had access, if the infrastructure was correctly constructed, to free sea water (although as we have seen in the case of Brighton, the quality of this water was often very poor). However, this meant that they were dependent on the fortunes of the town more generally and as its fortunes waxed and waned so did the fortunes of the aquarium and, while the general attractions of a seaside town created a large pool of potential paying customers, they also were a source of competition. Zoos are by their nature open-air institutions whereas aquariums (excepting a few tanks for marine mammals) must be indoors which means, unless the building housing them is exceptionally large, that there is very little space for anything except the tanks. And this is another reason for the massive capital that had to be expended on some projects. Paradoxically, the space created was for things such as roller-skating rinks, that had nothing to do with the fish. But only by providing those things could the aquarium company persuade people to come more than once.

Finally, fish are very hard to look after and keep healthy, and even harder to source and transport in good condition. The mortality rate in Victorian aquariums was horrific and the cost of keeping the fish alive, whether via Lloyd's constant circulation system or Saville-Kent's aeration method, was substantial when compared with finding meat for the lions or hay for the zebras. And the evidence shows that, given the limitations of the science and available technologies, Victorian zookeepers were good at their job when this is measured by the survival of their animals. This was partly because exotic animal keeping had a long history and a body of knowledge had been built up over the years and partly because most of the large animals kept in Victorian zoos were quite hardy and adaptable to their alien conditions. Fish, on the other hand, are delicate and cannot adapt to even quite small changes in water temperature or composition. So, the challenges of keeping an aquarium financially, technically and with a view to animal welfare were very significantly greater than the challenges of maintaining a zoo. And very little was known about the habits and physiology of fish compared with what was known about land animals and birds – which is still the case.

Nevertheless, the dividends paid at Brighton during its early years were such as to blind people to the very real risk of failure that investment in an aquarium project entailed. As Lee Jackson has pointed out, the potential financial reward meant that:

> The promotional benefits of claiming to build an aquarium were clearly more important than the costly practicalities of actually doing so.[4]

The pages of local newspapers from the 1870s are filled with accounts of shareholders' meetings of numerous aquarium projects and the various legal processes needed to establish and build an aquarium. Many – such as Tynemouth, Margate, Llandudno and Rhyl – failed before they were even begun, although the Rhyl Winter Gardens, Aquarium, Land and Building Company did raise the capital to construct a magnificent building that opened in 1876, containing a concert hall and a roller-skating rink which, by 1889, had lanes marked out with coloured lights. At Margate, the Margate Skating Rink and Aquarium Company raised the capital to build a successful skating rink, swimming bath, concert hall and restaurant in 1877. After two decades the business was sufficiently solvent for the planned aquarium finally to be added but, unfortunately, the seafront building was of a flimsy single brick and galvanised iron construction and was entirely washed away in a great storm on 29 and 30 November 1897.[5] The Sutton Coldfield Aquarium and Skating Company also got its rink up and running in 1879 but the promised aquarium never appeared. Others started but then failed. None had the enduring success of Brighton or the Crystal Palace.

By the end of the decade the ultimate expression of the aquarium as investment comes into view: The Westminster Aquarium, or Royal

---

4   Jackson, L., *Palaces of Pleasure* (New Haven, CT: Yale University Press, 2019), pp. 208–9.
5   See Howe, M., *Old Rhyl from the 1850s to 1910* (Rhyl: Gwasg Helygain, 2000) and Barker, N., Brodie, A. and Dermott, N., *Margate's Seaside Heritage* (Swindon: English Heritage, 2007).

Aquarium and Summer and Winter Gardens. This was initiated to address a gap in the market as there was no aquarium in central London. It hosted all manner of entertainments from swimming galas through circus acts and concerts to cat shows and "the first ever gorilla in Europe" (borrowed from Berlin Aquarium) but, as an aquarium, it was not a great success. This was despite the enormous infrastructure installed with reservoirs containing 700,000 gallons of water to service the tanks, using a continuous circulation system installed under the supervision of William Lloyd himself. This was powered by patent rotatory pumps from Leete, Edwards & Norman. The tanks themselves contained only 150,000 gallons or just over 20 per cent of the total water kept in the building. There were 33 tanks, which contained both marine and freshwater fish. *The Era* (16 January 1876) reported that the pipes were all made of vulcanite to prevent the contamination of the water by metal (another Lloyd trademark feature). In order to achieve continuous circulation, the water had to travel through three miles of pipes. This distance depleted the oxygen and the many joints required caused constant minor leaks. The proprietors – who always tried for the best: they had John Everett Millais as head of the art department and Arthur Sullivan conducting concerts – called in William Saville-Kent who modified the Lloyd system to his own continuous aeration system, which maintained reasonable water clarity and was much less expensive to operate. Even so, the showpiece, the "largest salt water tank in the world", appears never to have been filled for the purpose of the displaying fish. The building was opened in a very incomplete state and the *New York Times* (12 February 1876) commented that "the tanks for the much vaunted aquarium contained neither water nor fish". By the middle of 1876, Saville-Kent had stocked most of the tanks and while he admitted that it was regrettable to have opened an aquarium without fish he pointed out that:

> It was not until the succeeding summer that the entire series of tanks were placed in a state of perfect order and fitted for the reception of aquatic animals and plants.[6]

---

6   Saville-Kent, W., *Official Handbook to the Royal Aquarium, Westminster* (London: Charles Dickens & Evans, 1877), p. 3.

6 The public aquariums 2

These animals and plants constituted a collection:

> which, for size, number and variety has not been equalled at any existing inland aquatic exhibition. Many of these ... are now for the first time exhibited to the British public in a captive state.[7]

It was indeed a large and varied collection and Saville-Kent's *Official Handbook to the Royal Aquarium Westminster* lists 121 different plants and animals. Notice that Saville-Kent makes a distinction between inland and coastal aquariums; this is important to understanding the significance of at least some of the Westminster exhibition. For example, the Royal Aquarium had mackerel. They are common enough fish but they are very delicate and almost always die when they are transported. Coastal aquariums could fish them out of the sea and take them direct to the tank but an inland aquarium could not. So, to see free-swimming mackerel in the middle of the metropolis was something that needed to be explained to be fully appreciated and speaks to Saville-Kent's aspirations for Westminster as a site of science and education. Saville-Kent also set up an Aquatic Natural History Museum adjacent to the aquarium.

But scientific mission notwithstanding, the Royal Aquarium tried to maximise the entertainment (and thus the commercial) value of the fish by sequencing feeding times with the various shows in the music hall and elsewhere:

> The feeding of the animals having always proved an especial attraction to visitors, it has been arranged that this shall take place at a fixed hour, commencing at Tank 1 immediately after the afternoon entertainment.[8]

It was obviously a dramatic event but the hand of Saville-Kent and, I hope, the directors' sense of decorum, kept it within the bounds of good

---

7   Saville-Kent, W., *Official Handbook to the Royal Aquarium, Westminster* (London: Charles Dickens & Evans, 1877), p. 3.
8   Saville-Kent, W., *Official Handbook to the Royal Aquarium, Westminster* (London: Charles Dickens & Evans, 1877), p. 4.

taste. This could not be said of events across the Channel where the *Manchester Courier* (29 August 1874) reported that:

> The inhabitants of Havre [*sic*] seem to be making a bad use of their fine aquarium by setting octopus and conger eels to fight.

But "the Aq", as it was known, soon became an aquarium with hardly any fish. As Errol Sherson remembered:

> The fish, though few in number, were on view for some time; in fact, I think some lingered on to the very end twenty-seven years later in 1903 – but I have always wondered whether anyone went to look at them or if the water was ever changed!⁹

One attempt to develop a star animal failed when a whale that had been shipped in from Labrador died "of a cold" ten days after it was delivered. It had been transported on an open deck and constantly soused with sea water. It had been insured during transit but not after delivery so the aquarium took a loss of £1,000 (*Illustrated London News*, 20 October 1877). It was alleged that the whale had been ill-treated (one would have thought that the manner of capturing it and transporting it alone constituted that) but *The Times* (3 October 1877) took a more charitable view:

> When once the animal was safely deposited in the tank its surroundings were fully as favourable as those of most other creatures when deprived of their natural liberty. The supposed marks of ill-usage on the body were the consequence of the eels in the tank having after its death nibbled the edges of its fins.

We hope for the whale's sake that it was *post mortem* that the nibbling happened. The illustrations show that it was a very snug fit for the tank so it would not have been able to swim away to avoid the eels. Although the engraving *The Dead Whale at the Royal Aquarium* published in the

---

9   Sherson, E., *London's Lost Theatres of the Nineteenth Century*, (London: John Lane, 1925), p. 297.

6 The public aquariums 2

Illustrated London News (6 October 1877) shows no sign of damage, the fact that the management was still taking the aquarium seriously is shown by the fact that they retained the reputable marine scientist Henry Lee to investigate the whale's fate so that any whales they might be able to buy in the future would fare better. Alarmingly, the whale was supplied by Coup, whom we have met earlier, and with whom the management of the aquarium had contracted to buy more via William Hunt ("The Great Farini") who organised much of the entertainment at the aquarium. Next year another four arrived, of which three survived the transatlantic journey in boxes packed with seaweed. One of these went to Westminster and *The Graphic* (8 June 1878) published an engraving showing the whale being captured in "Farini's Whale Bay, Labrador", transported by ship, carried into the aquarium and finally in its tank. The whale became part of Farini's show and there was a charge of 1 shilling extra to see it.

Although the Aq ultimately failed as an aquarium, it enjoyed significant success for a time as a music hall, weathering several storms caused by the alleged indecency of its acts and the loose morals of its clientele.[10] Although as *The Times* (19 October 1889) sensibly pointed out:

> It is not easy to see how women of a certain class or those who seek their company can be excluded from a place of public entertainment if they pay for their entrance with the rest of the public and behave like the rest of the public when they get inside.

The failure of the scientific mission was the result of the relentlessly simple demands of the profit-driven business model and the unrealistic expectation created by the high dividends paid at Brighton. As *The Era* (12 September 1885) put it:

---

10 See, for example, Davis, T.C., "Sex in Public Places: The Zaeo Aquarium Scandal and the Victorian Moral Majority", *Theatre History Studies*, 9 (1990), pp. 1–13. See also Purdon, J., "At the Aq", *Junket*, 2 (9 January 2012), thejunket.org/2012/01/issue-two/at-the-aq (9 January 2012). For some account of the kinds of swimming displays put on, see Ray, D. and Roberts, M., *Swimming Communities in Victorian England* (London: Palgrave, 2019).

In all cases where a corporate body provides the capital, a dividend is expected, and to pay this it has been found necessary to abandon the lofty ideals propounded in the prospectus issued to the public when the Aquarium was in its embryonic stage.

In the 1879 edition of his *Dickens's Dictionary of London*, Charles Dickens Jr summed it up very accurately:

> Unfortunately, the desire of the directors to obtain an immediate return for the large sums invested in the undertaking unduly precipitated the beginning of the campaign. Not only were the plans of the managers in an inchoate state, but the tanks without fish became a standing joke, and the dissension among the discontented proprietors further tended to create distrust of the enterprise in the public mind. After some time, the tanks were filled and energetic management provided attractive entertainments of a superior music hall type.[11]

My guess is also that the Crystal Palace soaked up all the demand there was for an aquarium in London, even though it was a way out of the city.[12] So if you really wanted to see fish you went there and came to the Aq for racier pleasures.[13]

The Manchester Aquarium was the first of the public aquariums and we may now look in some detail at its short life as this is representative, with varying local detail, of the life of many of the

---

11 Dickens, C., *Dickens's Dictionary of London* (London, 1879).
12 The London Aquarium Company applied to open an aquarium on the Victorian Embankment in 1873 but was refused planning permission by the Metropolitan Board of Works on the grounds of the enormous quantity of earth that would need to be removed and then disposed of. The East London Aquarium, Menagerie, and Wax Work Exhibition opened in 1875. This was a place of general entertainment aimed at a working-class public and charging only 1d admission. It appears only to have had a few tanks including some seals. It burned down in 1884.
13 The only relatively complete account of the Royal Aquarium I have found is Munro, J.M., *The Royal Aquarium: Failure of a Victorian Compromise* (Beirut: American University of Beirut, 1971).

aquariums set up during the 1870s and, as we shall see, its failure should have warned investors in the Royal Aquarium to lower their expectations.

The Manchester Aquarium was first projected in 1872 and was intended to be built on a grand scale. In fact, it was to be twice the size of any existing aquarium. *The Architect* (1 June 1872) noted that:

> it was to be a building to exceed in magnitude the great aquarium at Sydenham.

The building was designed by James Sherwin who, perhaps ominously, went bankrupt in 1873. In addition to the main aquarium building (which was planned to contain 68 tanks), the original design included a winter garden with conservatories and marine aviaries. As *The Architect* put it, capturing very well the optimism and ambition of the times, it was to be:

> an intellectual and pleasing retreat in the colder months of the year.

Which in Manchester in the later nineteenth century was most of them.

From the earliest days, the mission of the Manchester (or Whalley Range) Aquarium was educational. It set its face against the gaudier attractions of Brighton or the Crystal Palace and reverted to the kind of statements that were made about the Fish House some 20 years earlier. This mission was maintained, and it is hard not to conclude that rigorous adherence to the purpose of the moral and intellectual improvement of the people was what caused the problems that dogged it before it was even completed. The aquarium company was set up on the basis that it would raise funds sufficient to complete the entire project. But it failed to attract enough investors, so only the main aquarium was completed and this cost £8,000 when only £5,000 had been subscribed. So, it looks as if the cautious investors of the north-west considered that a highly capitalised project which depended only on education for its income would not offer the same returns as the more diversified business at Brighton.

Nevertheless, the aquarium opened on 21 May 1874 and this event was recorded in the *Manchester Courier* the next day. The paper reported the high ambition of the aquarium's mission:

> if there was a "strange fish" in any part of the world they should have him in the Manchester Aquarium (applause). They should endeavour to make the Aquarium the means of the healthy moral improvement of the people, and the means of instruction in a very important science.

Frank Buckland, complete with crocodile and fake mermaid, made a speech and in this he reiterated and extended the purpose of a large aquarium:

1. It was to encourage observation of nature.
2. It was to show the beauty and design of creation.
3. It was to increase the people's food.

Before the response from the chair of another project group, The Southport Aquarium and Winter Garden Company, Buckland also got the Manchester collection going by donating some salmon, some trout and some golden tench from Germany. The next day the *Manchester Courier* published a letter of support from the Bishop of Manchester who approved of this edifying new feature of the city.

The board employed William Saville-Kent as "Curator and Naturalist" and he supervised the ambitious engineering project required to set things up. This involved steam engines moving 260,000 gallons of sea water (brought in from Blackpool) at the rate of 80 gallons a minute to achieve total circulation and filtration every 40 hours. Saville-Kent was fully in harmony with the scientific and educational mission of the aquarium and used it to pursue his own research. He addressed and solved the previously intractable problem of keeping large seaweed plants in good health in tanks, brought in whitebait from Colwyn Bay and, over 18 months, was able to show how they became herring; he observed the development of lobsters from eggs to young adult and demonstrated that fish like herring and dogfish needed to be kept in dim light at night to preserve their ability to navigate. It will be noted that all of these experiments (except perhaps the one concerning seaweed) had practical implications for the fishing and seafood industry and, in this, Saville-Kent was following in the footsteps of Frank Buckland (who may well have recommended him for this post) by using the aquarium as a motor for the increase of food production from the sea. Saville-Kent also wrote about a rare snake the aquarium had acquired. This was the Congo snake, *Muroenopsis tridactula*. It was thought to be poisonous but this, he wrote, "was an injustice to the poor creature as great as when applied by our own benighted countrymen to the harmless Newt or Triton of English ponds and streams".[14] However, what appear to be mainly personal reasons meant that Saville-Kent's tenure at the Manchester Aquarium was short-lived and by February 1876 he had resigned and moved back to London, partly to take on the role of curator and naturalist at Westminster and partly to act as consulting naturalist to the new aquarium at Great Yarmouth on the design of which he had previously advised and where William Lloyd was also a paid adviser. But he did leave the aquarium a great gift: his excellent and comprehensive *Official Guidebook to the Manchester Aquarium* which was published in 1875 and was a model of its kind. In its warm review of this publication, *Nature* nevertheless felt constrained to point out that:

> The Manchester Aquarium ... has attained considerable success, although we believe it has not quite realised the expectations formed of it by its original promoters.[15]

This was, to say the least, a tactful formulation. And some inklings of the problems Saville-Kent may have faced (as well as the problems all aquariums had to struggle with when it came to stocking their tanks) can be seen from a court case reported in the *Manchester Courier* (24 October 1874). The ex-curator Benjamin Hooper successfully sued a director of the company for £14 15s he claimed was owing to him for collecting fish. The director had been anxious that there were not

---

14 Saville-Kent, W., "A Rare Animal at the Manchester Aquarium", *Nature*, 12 (1875), p. 69.
15 Anon., "Official Guide-book to the Manchester Aquarium [review]", *Nature*, 13 (1875), p. 85.

enough fish in the tanks and had sent Hooper to get some. This he did but when they arrived 90 per cent were dead and so Hooper wasn't paid. However, he argued that he was contracted to put live fish onto a train not to deliver them alive and, perhaps surprisingly, this submission won the day. One wonders how many times Saville-Kent had to deal with micro-managing directors.

In less than a month the *Manchester Courier* (19 June 1874) was able to report that the aquarium was keeping to its plan and had acquired:

> A specimen of the wolf or cat fish (*Anarhichas lupus* [this is the Atlantic wolf fish]) ... being especially interesting as the first example of the species yet to be kept and exhibited in a public aquarium.

Alas for the aquarium and the fish, this specimen died of fatigue from its journey in captivity and was stuffed to be exhibited alongside its replacement, which arrived in December 1874 (*Manchester Courier*, 26 December 1874).

On the same day this newspaper also reported on the company's Annual General Meeting, which was much more concerned with money troubles and the need to raise more revenue than with the fish. Two proposals were made. The first was to open between two and six o'clock on Sunday afternoons and the second was to open a bar. The first motion was lost by 26 votes to 24 and the second was lost also. Here we see the problems that beset, sooner or later, almost all Victorian institutions that opened to the public for educational or cultural purposes. The mass audience they required to make ends meet had only limited leisure time and so required work-friendly opening hours. This meant Sundays and this was not acceptable to the Sabbatarian lobby. Similarly, the idea that visitors might enjoy a glass of beer was also unacceptable and so the principles of profitable business constantly conflicted with the principles of moral control, and, on occasion, a majority of the shareholders were constrained to vote for their deeply held beliefs against their own financial interests. If the people were to be educated, they were to be educated on terms and at times determined by their social betters.

On 1 August 1874, an Extraordinary General Meeting tried another tack and authorised the directors to issue another 4,000 shares at £5 each with a liability set at 3d. This meeting also appears to have relaxed the precious stringency on Sunday opening as the record shows that in the last week of August of that year 5,000 people visited the aquarium with 1,780 visiting in four hours on Sunday, 22 August, and in the week before there were 4,400 visitors with 1,700 on Sunday. By December the aquarium was varying its attractions slightly with a band concert on Christmas Eve and a lecture on microscopic illustration. It was also now served by new bus routes and was actively managing its display. For example, the large octopus had been put in a more "visible" tank as it was always hiding in the rocks which adorned its old one. And this, we note, meant that this most intelligent and sensitive of living creatures was denied its natural behaviour. In addition, adjustments were made to ticketing so that annual and half-annual season tickets at reduced rates were put on sale.

Despite these efforts, the situation was becoming desperate by June 1876. The Annual General Meeting for that year immediately dissolved into acrimony because only 30 shareholders could attend, and it was angrily pointed out that this time suited the convenience of the board not the shareholders who promptly moved and carried a motion for adjournment. Even so, they did receive the report, which showed that, while admissions had brought in £3,158 10s and 2d, the aquarium had cost £3,497 19s 11d to run and had thus lost over £300 that year. Remedial actions included a decision to sell the land where the originally planned winter garden was to be and to introduce a ticket pricing structure of 5s annual tickets with weekdays at 1s and Saturdays at 6d. They had also lost their curator. I can find no evidence that anyone of similar experience or ability to Saville-Kent was appointed and it is hard to think who there would have been who had both the requisite engineering and ichthyological experience and knowledge. The board thus had the challenge of running the aquarium, that most technical and delicate of institutions, without the benefit of a resident scientist who also understood how the place worked.

The meeting reconvened on 26 July 1876 and the directors came out swinging. They blamed the shareholders for the current crisis in the business as they had not supported the aquarium and had spread tales that it was a failure (*Manchester Courier*, 1 July 1876). It was proposed to set up 200 mortgage debentures at £300 each. Dividends had been paid at ten per cent in one year only and this was in spite of the fact many shareholders had put their money in four years earlier. The shareholders then set about the directors and complained that they had taken their fees even as the business failed and should return them. The chair, perhaps unwisely given the tone of the meeting, pointed out that the venture had been a success for the first six months since when it had suffered a series of mishaps.

The shareholders now had the bit between their teeth and attacked the directors' pricing strategy. It was pointed out that doubling the price of admission to 1s rather than cutting expenses (such as directors' fees) had been a poor plan that had resulted only in reducing visitor numbers by half. As one shareholder put it:

> To charge a shilling to see a few fish was as good as putting up a notice "No Admission" (Hear, hear).

A motion to bring the price back to the original 6d was carried and a new tariff of 6d for weekdays, 3d on Saturday and Sunday and 1s on Friday, with children's tickets half price, was carried. Notice how the weekend was cheaper to cater for the working class and that there remained one expensive day so that the carriage trade did not have to mix with the people at large. But the comment from the shareholder quoted above is worth unpacking as it demonstrates an excellent grasp both of the market (which depended on a mass public) and the shortcomings of the product: "a few fish", however interesting, did not stimulate repeat business. As the manager of the Morecambe Aquarium remarked when it closed in 1896 to make room for more entertainment space, "people would rather see a juggler than an uncooked lobster". By early November, admission prices were halved again to try to stimulate demand.

As 1877 dawned the aquarium tried to diversify its business. An arctic exhibition based on the *Discovery* expedition was staged in January and this included panoramic light projections and a performance by "Mr Greenwood the well-known nautical ballad singer" (*Manchester Courier*, 1 January 1877) and, in the next month, a magic lantern show based on the Prince of Wales' recent visit to India was staged three nights per week and included anecdotes, songs and music (*Manchester Courier*, 27 February 1877).

However, all this was to no avail and in July 1877 the secretary of the company wrote to shareholders proposing that the company be wound up. The admission charge reduction to 3d, which appeared to be the only price at which people would come, had reduced gate receipts to £946 3s 6d and this simply was not a viable level for the business to keep going. It was pointed out that, had the building been in a proper state of repair, a dividend of 1.75 per cent would have been payable from a small profit. This would have been cold comfort to the long-suffering shareholders who had seen nothing for years and who had bought into the business on the expectation of returns like those payable at Brighton. This point about the repair of the building probably refers to two problems that had dogged the building since its construction. The first was persistent leakage of sea water from the tanks. The second was that the enormous quantity of water required had put stress on the floors, which were not sufficiently supported to hold the weight.

The secretary's letter, which was reported in the *Manchester Courier* (2 July 1877), did set out a plan by which the aquarium might be viable and admitted that the current business model offered no hope of success. Success might have been achieved by reversion to the original plan. The building should have been extended and placed in a better state of repair and efficiency, generally improved ("beautified") and freed from debt. The land should have been enclosed and the winter garden established together with a museum and herbarium. The letter not only addressed the shortcomings of the incomplete estate but also those of the business model and proposed that the aquarium might have succeeded if it had had facilities to mount technical exhibitions "and other attractions". It concluded that if all these things had been done or were to be done then "there is good reason to believe that financial success would be obtained". A final attempt was made to raise another £8,000 in capital but this attracted only £2,500 (which given the state of the company is a surprisingly large amount).

The shareholders' meeting reported in the *Manchester Courier* only a few days later (5 July 1877) agreed to wind up the company with debts of £6,552 5d. It was noted that by this time visitor numbers had dropped again and revenue from admissions was averaging only £128 per month. An appeal had been made to the city council to take on the facility as a municipal facility but this had been rejected. The position was sadly summed up:

> The aquarium was opened to see if such a place was wanted by the inhabitants of Manchester, and as it was found that they did not want an aquarium it was no use keeping it open.

This is perhaps only partly true as a major motivation for the enterprise was speculative profit and, in addition, what were clearly poor decisions and indecisiveness about ticket prices and opening times did not help to establish a regular audience. Finally, the main experience on offer, "to see a few fish", was simply not attractive enough to establish the aquarium as a frequent resort for enough people. And it is probably true that had sufficient capital been raised in the first place to complete the scheme as planned to a high quality then it may well have succeeded for a few more years at least. The Victorian enthusiasm for exhibitions of all kinds might alone have underpinned less-profitable operations like the fish. One shareholder made a passionate statement:

> It was also opined, by Mr Brittain, that to let a paltry sum of £6000 or £8000 to stand in the way of education and science in the city of Manchester would be a thing of which he, for one, should always feel ashamed.

This sentiment clearly held some attraction as the *Manchester Courier* (16 July 1877) published a soul-searching piece on the collapse of the aquarium, which pointed out that the aquarium was "only an aquarium" and in spite of recent attempts to hold exhibitions and other events it was therefore quite different in character from others elsewhere, with the correct implication that success could not depend merely on a display of fish, however educational this might be. In addition, this piece identified the chief supporters and paying customers as artisans. So "a great utility in working class improvement" had been lost and it would be a shame if the council did not rescue the project and, perhaps, link it with the city art gallery in some way. This plea also fell on deaf ears.

The building was put up for auction but attracted a maximum bid of only £6,750. The auctioneer then suspended the sale for ten minutes to let people have "a little silent reflection on the matter" (*Manchester Courier*, 2 August 1877). This brought no enlightenment, so the sale was abandoned. Shortly afterwards the building was sold by private treaty to St Bede's School and still forms part of that institution. It is apparently even now possible to trace the warping of the floorboards caused by the leakage and the weight of the sea water in the tanks.

The story of the Manchester Aquarium can be seen repeated throughout the 1870s and the two main reasons for failure of these institutions later in the decade are fully apparent: lack of capital (or rather over-ambitious capitalisation given the likely returns) and the sheer difficulty and expense of maintaining a large volume of sea water in tanks. Manchester was actually quite modestly planned compared with the projects that followed it: Scarborough cost £70,000 to build with a subsequent investment of £110,000; Morecambe and Rhyl were costed at £35,000 and £35,080 respectively; Westminster a massive £178,000 (reflecting the cost of prime building land in central London); and Great Yarmouth £38,000.

Unlike its ill-fated predecessor in Manchester, the Marine Aquarium (Scarborough) Company managed to raise by subscription all the capital it required for the construction of its ambitious aquarium, which was designed by Eugenius Birch (who also built Brighton) and this opened in 1877 although it had been planned to be finished in 1876. The *York Herald* (1 January 1877) reported on the progress of "the finest aquarium in the country" and perhaps optimistically added that "wherever aquariums have been established they have never become a worn out novelty". The aquarium sought to exploit its seaside location by connecting directly to the sea and having the water wash straight in. The final cost was £120,000 and the *York Herald* (21 May 1877) reported that a number of investors also had financial interests in Brighton. The building of the aquarium was a natural progression for Scarborough, which was one of the first seaside resorts where, as early as 1788, shops selling seashells were operating and where, by 1825, there were at least four specialist dealers in shells and fossils.[16]

The aquarium opened on 19 May 1877 and held a special event for locals which attracted 3,000 people. The staff wore naval uniforms with the word "Aquarium" picked out in gold on their hats and the red-and-white SA monogram on their blue shirts. The event was a success with a particular attraction being a display of puffins and guillemots who could be seen, through the glass of their tank, happily pursuing fish. But an immediate problem presented itself, which was that the aquarium did not serve alcohol and people who nipped outside for a beer found that they couldn't get back in. This was an issue partly symptomatic of the problem for many Victorian institutions where the provision of wholesome leisure for the working class clashed with the explicit social control of an alcohol-free (and therefore, it appears, fun-free) zone. But in Scarborough in particular the cultivation of its brand as an upmarket resort catering for the middle class and above sat uneasily with the development of a facility such as an aquarium that could only pay on the basis of a mass audience. Thus, the aquarium, which covered two-and-a-half acres, had a very unobtrusive entrance to hide the proletarian visitors from the delicate view of more refined holiday-makers and day-trippers.

There were 26 tanks, one holding 75,000 gallons of water, and a seal pond as well as other facilities such as a restaurant and reading room. There were some interesting fish including a sturgeon and that indispensable attraction, an octopus. But the requisite visitor numbers and revenue never materialised, even though new bus routes were developed specifically to serve the aquarium. By 1881 the aquarium company was bankrupt. The *York Herald* (26 May 1883) reported that a meeting of unsecured creditors was offered part payment in shares (which can't have been an attractive offer) and told that the aquarium would reopen on 1 July 1883 under the management of Captain Riddle of Drury Lane Theatre. Here we see clearly that the business model of the aquarium was shifting in favour of entertainment but Captain

---

16   For a collection of historical anecdotes about Scarborough aquarium and its successors, see
storiesfromscarborough.wordpress.com/tag/scarborough-aquarium/

6 The public aquariums 2

Riddle could not solve the puzzle. On 25 May 1886 the premises were put up for auction. It was sold for 50 per cent of its value but, even then, ongoing revenue was insufficient to meet expenses or even interest payments. The company secretary sadly admitted that:

> As an aquarium, it was, we regret to say, a complete failure, and even with the introduction of varied amusements could not be made a success.

(*The York Herald*, 25 May 1886)

The new owners tried to sell it for £18,000 and then dropped the price to £10,000. They finally recouped some money when the building was sold to the impressarios Starkey and Morgan for a mere £4,000 with an additional £1,150 for fittings and was turned into The People's Palace, which was a music hall specialising in sensational acts. The scale of the aquarium can be measured by the fact that some of the tanks became shops, and its past was dimly recognised in an alligator tank and an exhibition of endurance swimming by Captain Webb in the biggest tank (which was 26 feet square). In this guise it carried on with varying success until it was demolished and turned into a carpark in 1968.

What Scarborough shows is that an aquarium without significant other attractions could not make money and also, arguably, that location was crucial and that the free-spending and jubilantly working-class Londoners who carried Brighton Aquarium to such heights of profitability for a few years were simply not available in the poorer north and in a resort that prided itself on its gentility.

A not-dissimilar tale played out at Great Yarmouth, although here it was the availability of other entertainments not the lack of them that caused the demise of the aquarium. The original plan developed by the Great Yarmouth and Eastern Counties Aquarium Company required initial capital of £50,000 but only £25,000 was raised. Nevertheless, the *Illustrated London News* (9 October 1875) reported a decision to go ahead with the erection of an aquarium and skating rink at a cost of £30,000 with the land donated by the town council. A year later (9 and 16 September 1876) the same magazine reported that a magnificent aquarium with massive tanks was starting to appear although at this point it was still only partly built. Eventually the work was completed, the tanks were stocked (although there is hardly any evidence of what was in them or how they were managed) and the skating rink, reading room and café, all by now the familiar accoutrements of a large aquarium building, were up and running.

The company had employed William Saville-Kent as the manager and William Lloyd as a paid adviser. But this arrangement did not last and Saville-Kent soon departed to try to set up his marine research station on Jersey. In a letter to his old friend and patron, Richard Owen Lloyd summed up his views on Saville-Kent's capabilities as an aquarist bluntly but not entirely unkindly:

> He made a dreadful mess of the Westminster Aquarium after I voluntarily left it, and then he blundered in the Regent's Park one. Since he left the British Museum for aquarium work in 1871, he has been in five public aquaria and has done no good, and has not made any reputation, in any. He is now in Jersey trying to get up an Aquarium costing £3,000 or £5,000 in £1 shares, but I fear for his success, and I think he has mistaken his vocation, as it wants a man to be more than a mere naturalist to be a good aquarium maker and manager. He should never have left the British Museum for an occupation requiring much and dirty hard work, and where kid gloves and chimney pot hats have no place. I have helped him in a most disinterested manner, with money and work, and influence, but he seems to have a curious facility for separating himself from his friends.[17]

Perhaps one senses here a thing that is rare in Lloyd, who appears to have been of a most friendly and generous character, and that is the resentment felt by a self-made man who at one time could not afford even a few pennies to get into the Fish House or buy a book towards a middle-class gentleman who had had access to every educational and social opportunity and yet has made little of them.[18] The fact is that

---

17 Parlouraquariums.org.uk., *Parlour Aquariums.*
18 Saville-Kent had a difficult start in life as he was the brother of Constance Kent, the notorious Road Hill House murderer, who is the subject of Kate

6 The public aquariums 2

Saville-Kent was a fine ichthyologist and marine scientist but only a moderately successful aquarist.

It is clear, however, that the real attraction of this building was its music hall–style entertainments and *The Era* (26 December 1880) reported not only the renewal of the drama licence but also the order to wind up the company, adding that the focus of the business had changed dramatically and that the aquarium "is becoming a sort of Palace of Varieties". This stealthy modification of the offer created problems as, when the liquidator of the company applied for the renewal of the licence in 1881, he was turned down on the grounds that there were already enough theatres in Great Yarmouth. His defence, as reported in *The Era* (1 March 1881), is instructive in showing how a Victorian business might contextualise an aquarium and the experience it offered when this wasn't directed by a strong scientific or educational mission:

> It was built entirely for an Aquarium; but he would ask the magistrates if it was really an Aquarium at the present time. Where were the fish or anything pertaining to an aquarium?

A good question. Where were the missing fish? One can only assume all dead.

The building was auctioned but not sold mainly because a restrictive covenant in the lease meant that the council could take possession of the building if, by 1883, it was not completed to its original specification.

Eventually a buyer was found and the premises reopened as the Royal Aquarium under different management and with a £10,000 extension containing extra attractions, including a space in which celebrity lecturers, including on one occasion Oscar Wilde, could hold forth. Although the prime business was now entertainment, the aquarium nonetheless persisted and an advertisement for the theatre in the *Eastern Daily Press* (4 September 1883) included the enticement

---

Summerscale's best-selling book *The Suspicions of Mr Whicher* (London: Bloomsbury, 2008). It may be this that caused him to change his surname twice.

that there were tanks containing "thousands of marine and fresh water fishes". These were arranged on the side aisles of the main auditorium so while you were enjoying the music hall acts or boning up on greenery-yallery aesthetics with Oscar, you were surrounded by fish that watched you as you watched the stage. This experience was, as far as I can tell, unique in England and the logical conclusion to the search for a business model in which an aquarium and a music hall could sit side by side. The building is still in Great Yarmouth and, apparently, most of the tanks are also still there, hidden behind the walls.

While all this was happening on the east coast, on the west coast the growing resort of Southport was also developing an aquarium, which opened on 16 September 1874 and cost £100,000 to build. Two thousand people turned out to hear the band and also the opening speech by Frank Buckland, which, judging by the contemporary reports, was the same one that he had given at Manchester.

The enterprise was a speculative venture by the Southport Aquarium and Winter Gardens Company and was an enormous building covering nine acres and occupying a frontage of 1,100 feet on the town's most prestigious street.[19] There were 25 sea-water tanks, including one 63 feet long and 14 feet wide – which contained a shoal of cod and a monkfish of enormous proportions – and several freshwater tanks including table tanks. Water was drawn direct from the sea and piped 1,400 yards into the storage reservoir. From there, 150,000 gallons were in constant circulation, supplemented by a compressed-air aeration system. The collection was extensive and included an octopus. As with the original aquariums, much attention was paid to sea anemones and molluscs and there was a seal tank containing five seals who divided their time between it and the seal pond in the garden at the front.[20] Real efforts were made at the beginning to establish the aquarium as a serious rival to those at Brighton and the Crystal Palace

---

19 This was Lord Street, one of the premier shopping streets in the north of England and built sufficiently wide to allow a coach and four to perform a complete circle. During his years of exile previous to his return to France, the future Napoleon III spent time in Southport and it is said locally the glorious proportions of Lord Street gave him the idea for the boulevards of the re-engineered Paris.
20 Rance, C. de, "The Southport Aquarium", *Nature*, 11 (1875), pp. 393–4.

and the curator M.H. Read was especially keen to stress that he had sourced fish not found in the other established aquariums, including conger eels and wrasse, and the *Illustrated London News* (3 October 1874) reported that visitors to Southport could view the "largest sea perch ever caught". On 28 August 1878 the aquarium acquired two sturgeon both over seven feet long. The resident scientist was C.L. Jackson, whom J.E. Taylor described as "a careful and diligent naturalist".[21]

The building also contained other attractions, including displays of pictures of fish, a theatre, at times a menagerie, consisting largely of monkeys, and an opera house. Despite all this it was never a great success, and the company went bankrupt in 1898. Yet while Southport and Scarborough struggled on trying to find the right mix for profitability over several years, we might consider them great successes when compared with Tynemouth aquarium and winter garden. This was projected at a cost of £80,000 and opened on 28 August 1878, "clearly far from complete" according to the *Sunderland Daily Echo* (29 August 1878). The parent company was wound up less than six months later, on 22 March 1879. The business struggled on and by 1 August 1879 it was announced that it was trying to make ends meet by offering a reduced admissions price:

> to meet the means of the large body of excursionists now visiting the town and that "people's days" will be limited no longer to Monday only, but after the current week the price of admission will be reduced to sixpence at least four days in the week.

(*Sunderland Daily Echo*, 1 August 1879)

This strategy also failed and the aquarium closed on 15 November 1879, with the company owing nearly £17,000. On 14 January 1880 the building was sold for £27,150 (on a valuation of £90,000) to a new company that ran it as an exhibition hall, including in 1882 a very successful marine exhibition.

---

21   Taylor, J.E., *The Aquarium: Its Inhabitants, Structure, and Management* (London, Hardwicke and Bogue, 1876), p. 18.

The stories of Manchester, Scarborough, Southport, Tynemouth and, of course, Westminster can be traced in most aquarium projects throughout the decade.

Thus far this chapter has concentrated entirely on English and Welsh aquarium projects and the only Irish aquarium was dealt with in an earlier chapter. Some interest in an aquarium in Glasgow was reported in the *Edinburgh Evening News* (19 September 1878) but this came to nothing. The first aquarium in Scotland was the substantial building at Rothesay that was developed on land leased for a peppercorn rent by the Marquess of Bute and opened on 29 June 1876. The official guide, *Rothesay Royal Aquarium and Its Contents*, gives us a bracing view of the way in which a group of Victorian gentlemen might frame the scientific mission of such an impressive facility:

> If it should lead any, by becoming more intimately acquainted with the creatures around them, to recognize more fully the goodness and greatness of their Divine Creator, the Directors will feel recompensed for the outlay attending its [the guidebook] production.[22]

But, depending on the thought that God helps those who help themselves, they also included dozens of advertisements for local businesses at the beginning, end and throughout the book.

Rothesay was first projected in April 1873 and when it opened it could fairly describe itself as "the Brighton of Scotland". There was a freshwater saloon and a sea-water saloon, each with 11 tanks, placed on either side of a Grand Hall, which also contained four tanks. The Rothesay tanks were carefully designed to be leakproof and thus avoid the trouble and costs of losing water. The layout reverted to the earlier mode of aquarium display by using table tanks but the engineering was based on Lloyd's continuous circulation system. An innovative feature was a "Piscatorial Hospital" in which fish freshly arrived at the aquarium were given time to rest and acclimatise. The ticketing structure was quite different from any other aquarium with annual

---

22   Baker, E.A., *Rothesay Royal Aquarium and Its Contents* (Rothesay: The Directors of the Company, 1878).

tickets selling at a guinea for a family and half a guinea for single or monthly tickets at 7s 6d for families or half a crown single. This tells us a great deal about the expected class of visitors and the way in which the directors expected the aquarium to be used. Feeding times were regular and advertised, so they were also not immune to the notion that people just might find the collection entertaining as well as instructive. The collection was large and varied and, by 1878, 157 different kinds of creatures are identified in the guide. Fish came in from different sources: in July 1876 two seals were transported from Newfoundland. In December 1879 the aquarium received a cargo of specimens as a gift from the New York Aquarium. Michael's of Liverpool was the aquarium's official importer and in July 1878 delivered a cargo of turtles and alligators.

Rothesay was an initial success financially and in February 1879 the shareholders received the very respectable dividend of five per cent. It was also scientifically successful with the ova of the economically important herring being hatched there in 1884. But in 1888 the Fisheries Board of Scotland took over and placed the aquarium in the hands of a scientific team who worked on three projects: oyster hatching (in the old seal pond); herring hatching; and ova transportation.

In 1907 the Marquess of Bute bought the building and turned it into the Bute Museum, which still operates there today.

The *Edinburgh Evening News* reported the liquidation of the Edinburgh Theatre, Winter Garden and Aquarium Company on 17 March 1879. This had raised substantial capital for a theatre that opened in 1875 (and went bankrupt in 1877) but appears never to have got round to the aquarium end of the business (maybe an example of the use of the word "aquarium" to suck in investors?). It had, however, been an ambitious project that was aimed at developing "a well-stocked aquarium on a scale not inferior to those of Brighton and the Crystal Palace".[23] However, the corporation did establish an aquarium with

---

23 One of the very few traces of this would-be aquarium is a note in *Nature*, 11 (1875), p. 427, which mentions that it is being planned. See also Baird, G., *Edinburgh Theatres, Cinemas and Circuses, 1820–1963* (Edinburgh: privately printed by the author, 1964).

a "good selection of both sea water and fresh water fish and a seal tank" in the Circle in Waverley Market where, subsequently, the florist stalls were to be found. I wonder if they had in mind the similar operation at Covent Garden. This was proposed in 1877 and opened on 14 August 1878 and there is some evidence that it was still operating in 1883 as it acquired a wild-caught shark to accompany the wild-caught octopus and sturgeon it already had in its tanks. It is unclear when this institution (which also had a seal pond) closed. The only source for it I have found says that "it was done away with previous to 1890".[24]

There was some interest in an aquarium in Glasgow and a project meeting for a Polytechnic, Aquarium and Opera House was reported in the *Edinburgh Evening News* (19 September 1878), but this came to nothing. I suspect that by this time the prospect of raising the £80,000 required for the project, in spite of the promise of big dividends, just felt too risky. The *Edinburgh Evening News* (12 January 1897) reported a plan to open an aquarium and marine laboratory in Aberdeen, which also came to nothing.

If the Rothesay Aquarium enjoyed at least some success this could not be said for the Isle of Man's aquarium in Douglas. What by 1882 was characterised as "that gigantic failure, the Douglas Aquarium" (*The Era*, 17 June 1882) had begun most hopefully seven years earlier with a speech by the High Bailiff of Man to the would-be shareholders of the Douglas Aquarium & Baths Company Ltd. He said that there was:

> reason to expect that, owing to the ease with which specimens could be obtained, and the supply of water direct from the sea could be had, this aquarium would be greater success than any of the kind in the Kingdom had been.
>
> (*Isle of Man Times*, 18 September 1875)

This optimistic prognosis underestimated almost every challenge that aquariums faced but nonetheless the *Liverpool Mercury* announced the formation of an aquarium company on 21 October 1875 with the chair

---

24 Gemmell, P., *A Short History of the Old Green Markets and of the Waverley Market* (Edinburgh: Mould & Todd, 1906) p. 38.

of the board also being a director of the Manchester Aquarium (who should have known better by this time).

By November 1875 the plans for the building were on view and in addition to the usual aquarium tanks there were two innovative features: a spawning exhibition and a special tank for seahorse breeding. In addition, as at Rothesay, there was to be an acclimatisation facility that the *Isle of Man Times* (6 November 1875) saw as conferring a possible commercial advantage, pointing out that 90 per cent of fish died in transit or soon after. This stimulated great interest and more shares were applied for than could be allocated. In addition, a lively row broke out in the local paper between "Naturalist" who advocated that the aquarium should include an aviary, "ABC" who thought it shouldn't, and "Viator" who supported "Naturalist". So there was local participation. But 18 months later things were not going well. The *Isle of Man Times* (11 August 1877) reported a company meeting which showed that it had raised £25,000 in capital to fund a building estimated to cost £17,500 but was now faced with an estimate of £37,000 and the building was not even finished. It was proposed that the aquarium be abandoned and only the baths be built. It is interesting that already people were starting to realise that, if they were ever going to make any money, it would be from baths not an aquarium. A committee of investigation was set up to get to the bottom of things.

Only a week after this meeting, things took a serious turn for the worse when the Douglas Street Improvement Board repossessed the company's land by forfeit, citing breach of contract through the non-performance of the condition that an aquarium and baths would be built by 9 August 1877. However, the company managed to get some grace on this and the baths, "the finest in the Kingdom", finally opened (with no aquarium) on 25 August 1877 – an event that featured a swimming exhibition from the famous Captain Webb. But the baths, fine as they were, did not pull in the hoped-for custom and by December had raised revenue of only £186 8s, leaving the company with a perilously thin cash base of £154 14s 8d. To add to the directors' misery, the investigative committee report landed with a story of haphazard accounting, conflict of interest in tenders, unnecessary expenditure and general business naïvety. The shareholders' meeting reported in the *Isle of Man Times* (21 September 1878) was marked by "a good deal of recrimination", which I suspect is a very tactful formulation. In the third quarter of 1878 the baths took £405 10s against expenses of only £130 but this was not enough to turn things around and on 12 October 1878 the company went into liquidation with debts of between £4,500 and £5,000, including the architect's fee "which will be disputed" (*Isle of Man Times*, 12 October 1878). The bankruptcy case and final settlement paid off debts at 18 shillings in the pound and sold the building for £5,000. By June 1882 the baths had been demolished and replaced by a theatre. The shareholders got nothing back at all.

Finally, the aquarium at Aston Lower Grounds in Birmingham deserves to be mentioned as this was William Lloyd's last great work and one that he described in detail in the *Scientific American* in August 1879.[25] It was a magnificent building constructed in Byzantine style of multicoloured brick, with floors tiled by Minton, and above each of the larger tanks was a circular stained-glass panel by Bourne, each illustrating a fish-related quotation from Shakespeare. It had reservoirs containing 300,000 gallons of artificial sea water and asphalt-lined vulcanite pipes carried a continuous circulation system powered by six steam engines and two steam boilers. When Lloyd proudly described this marvellous building, which contained the fruits of the experience of his long and successful career, it had not been stocked and I can find no evidence that it was ever a success as an aquarium. In fact, Lloyd was constantly worried about his job there as he was concerned about the financial stability of the company. As things turned out, he was only employed for six months, during which time he had also fitted in some consultancy work for the Dohrn Aquarium in Naples. The Aston Aquarium was part of a larger entertainment complex that enjoyed over a quarter of a million visitors a year but, by 1886, it had closed and was replaced by a menagerie. In 1897 it was taken over by Aston Villa football club and converted into a gymnasium and offices. In 1981, this magnificent example of late Victorian architecture was demolished.[26] Lloyd's last job was some more consultancy for Crystal

---

25 Lloyd, W.A., "The Great Public Aquarium at Aston Lower Grounds, Birmingham", *Scientific American Supplement*, no. 189 (1879), pp. 3008–9.

Palace and for the publishers Cassel but on 27 July 1880 he died of a cerebral haemorrhage at the early age of 54.

---

26 An account of the systematic destruction of Victorian architecture may be found in Stamp, G., *Lost Victorian Britain* (London: Aurum, 2013).

7
Australia: an imperial case study

On 11 December 1867, the *Sydney Morning Herald*'s notices of custom import entries informed readers that Eastway & Son had taken delivery of 13 cases of aquariums from England. So, the way was now open for the citizens of Sydney and New South Wales more widely to encounter fish in their parlours in the way that these had been encountered in the old country in the past decade or so. The home aquarium boom had all but ended in Britain so it must have been a relief for an aquarium wholesaler to have been able to develop a new market for objects that were becoming increasingly difficult to shift. However, the arrival of these tanks was by no means the beginning of an aquarium craze in Australia and, as we shall see, the contours of this interest in aquariums followed very closely those that structured the encounter with fish in Britain, but several years later.[1]

The spread of aquariums through the empire was, in fact, not especially vigorous. We have already seen something of the first

---

1 For the non-Australian reader, it is worth pointing out that at this time each of the six Australian states were self-governing colonies in their own right and although the term Australia was used to refer to the continent, the political entity known as the Commonwealth of Australia did not come into being until Federation in 1901. In this chapter I will use "Australia" as a shorthand for "the Australian colonies" whenever I am not referring to individual states.

aquarium in India (in Madras, although there was a plan to have an aquarium at the Calcutta International Exhibition in 1883) and although several zoos had begun in South Africa before 1901 there appears to have been no aquarium there until 1931 when the East London Aquarium was opened. So the private aquarium of Egypt's Ismail Pasha, which was also referred to in an earlier chapter, was the only aquarium set up in a British-influenced part of Africa in the Victorian period. Canadians had to wait until 1956 for their first public aquarium (in Vancouver) although there was certainly home aquarium keeping in Canada long before that. New Zealand was the same, with the Otago aquarium opening in 1951 and Napier's in 1957 (although a display of goldfish had been set up in Les Mills' shoe shop there in 1954). And although Singapore plays an important role in zoo history, it too did not acquire an aquarium until the mid-twentieth century. Nor do there appear to have been any aquariums in the colonised Caribbean or Pacific islands, although there is now a very fine one on Noumea. The empire was, however, an unfailing source of ichthyological knowledge and specimens were sent back to London from almost every colony with major works, such as Francis Day's *The Fishes of India* (1875–88), Theodore Cantor's *Catalogue of Malayan Fishes* (1850) and Robert Playfair and Albert Günther's *The Fishes of Zanzibar* (1867) being presented to the scientific public and shifting the encounter with fish onto a global level.[2]

Australia appears, therefore, to have been unique in developing a fish and aquarium culture and this, like the corresponding culture in Britain, included home aquariums, public aquariums, an interest, both professional and amateur, in marine and littoral sciences, fish-related museum exhibitions and the expansion of the food economy to include pisciculture. In a way, the lack of a more pan-imperial fish culture is surprising as the expansion of natural history and ichthyology in Britain had been driven by the reports of military, naval and civilian scientists from all over the imperial world and museum collections in the home country were constantly expanded by specimens sent back from colonies, dominions and protectorates. My guess is that aquariums developed

---

2   See Saunders, B., *Discovery of Australia's Fishes* (Collingwood: CSIRO Publishing, 2012) for a full survey of early imperial ichthyology.

7 Australia: an imperial case study

more readily in Australia than elsewhere for three reasons: the first was the increasing wealth (from time to time), scale and sophistication of the Australian cities, which meant that audiences and markets were available to buy aquariums and to go to aquariums in sufficient numbers to support a fish-based culture; the second is the close connection between Australia and Britain, the relative ease of finding out what the latest crazes were in the old country and the desire to emulate them; the third is the growth of a genuinely Australian identity that, by the second half of the nineteenth century, had started to look at the sea and the beach as a distinctively Australian leisure space.

Although a big shipment of aquariums didn't arrive until 1867, we can trace an interest in home fish keeping and some pioneering endeavours earlier than that. In 1860 Samuel Hannaford's *Sea-Side and River-side Rambles in Victoria* had a picture of an aquarium as its frontispiece and the book was clearly intended as a distinctively Australian (Victorian) contribution to aquarium literature. Hannaford situates his book against others by stressing that it is user friendly and realistic about the kind of budgets someone wishing to establish a tank might have:

> Greatly as a taste for Marine Zoology has been diffused by the numerous beautifully illustrated and charming books which within the last few years have appeared from the English press, they are nearly all costly and many of them of too technical a nature to prove of real value to the masses.[3]

In his chapter on "The Aquarium", Hannaford advises setting up a tank using river water, seaweeds collected from St Kilda, Queenscliff, Warrnambool, Armstrong's Bay or Portland and stresses the importance of including rockwork in the tank so that the fish have somewhere to hide. He was an advocate of a natural-looking aquarium and advises his readers to construct simply and "not such fantastical grottoes as we have observed in some Cockney Aquariums". Hannaford is actually quite vague on the details of aquarium keeping and seems to

---

3   Hannaford, S., *Seaside and River Rambles in Victoria* (Geelong: Heath & Cordell, 1860), p. 69.

suggest that one should make one's own tank – which might be a clue that such things were not easily to be had in Melbourne at least.

We have already seen how, in the late 1850s, Australian newspapers had picked up on the English aquarium craze and expected their readers to have at least some idea of what they were talking about. By 1860 we see clear evidence of an aquarium craze in parts of Australia itself: many queries about suitable cement recipes, problems with offensive smells and milky water and recommendations for suitable aquatic plants begin to crop up in Australian newspapers and magazines. Eventually, the *Launceston Examiner* (27 July 1861) was able to point out that:

> As the Aquarium is now almost a domestic institution some of our readers will be glad to hear that they can make their own salt water artificially.

But while the aquarium was a familiar or at least a recognisable object in New South Wales, Tasmania and Victoria, things were different in South Australia. There, a writer in the *South Australian Register* (22 December 1860) was clearly worried that his readers might not know about them:

> In England scarcely a drawing-room is now found without an aquarium, which is a vase or tank of water filled with living plants and aquatic animals.

It appears that aquariums had been seen in Australia since the late 1850s. The *Moreton Bay Courier* (22 August 1856) contains the earliest reference I can find when it describes two aquariums on display in the Brisbane Botanic Gardens. However, these appear only to have contained aquatic plants and so this use of the word "aquarium" seems to be in the old sense of a water garden only. *Empire* (22 October 1857) mentions an aquarium on display at a charity bazaar in the Sydney Botanic Gardens and the same magazine (26 February 1858) also reports two aquariums at the Australian Horticultural and Agricultural Society exhibition. Although fish are not specifically mentioned in either of these notices, I suspect that these are more

7 Australia: an imperial case study

likely to have been stocked aquariums as the first article says that the aquarium was "a curiosity" and the second that "they were new to most of the visitors" and I don't see why attention should have been drawn to the novelty value of these objects if they were only water gardens. So, it may well be that these are references to the first genuine aquariums to be seen in Australia.

But the first unambiguous reference to a stocked aquarium comes in *Bell's Life in Sydney and Sporting Reviewer* (8 May 1858):

> There is on exhibition at Mr Lea's shop in Park Street, a very interesting specimen of art. It is termed an Aquarium, and the interior is tastefully formed in imitation of a submarine grotto … through which and around which a great number of little fish glide and disport themselves.

Notice that, at this point, the author still needs to introduce his readers to the word "Aquarium" (note the capitalisation). This aquarium was, one assumes, set up from patterns derived from an aquarium handbook as I doubt that the grotto aesthetic would have sprung fully formed from Mr Lea's head. Here we see, for the first time, that encounter with fish swimming at eye level that so captivated the Victorian public back in Europe being experienced by Australians in an Australian city.

Within a decade, the aquarium had become a familiar object and, although there are few anecdotal traces of the home aquarium, a reference in the *Sydney Morning Herald* (9 November 1867) to boys selling fish for aquariums on the street shows that they must have been reasonably common. And it was worth the while of the Sydney Glass Company to advertise in the *Sydney Morning Herald* (27 March 1868) that they were now making globes and home aquariums. On 31 December 1867, the *Geelong Advertiser* mentioned one in particular. This was being auctioned in aid of the Geelong Protestant Orphan Asylum:

> Mr Rigney's celebrated aquarium containing sea horses, snapper, gold and silver fish etc, will be offered for sale; an admirable opportunity is thus offered to those who take a delight in such

things; it would take them three months, and then, perhaps, they would not be able to get together such a collection.

One wonders if Mr Rigney's charity was also motivated by the realisation that keeping up a marine aquarium at home was not quite the simple thing that the aquarium books suggested. The heat of the Victorian summer must have made the task of keeping the water clear even more challenging and around this time we encounter other aquariums being sold off in charity bazaars and auctions so, as in England, it may well be that the early adopters were now falling by the wayside. A clue to how Australian aquarium keepers coped with the heat may be found in a reference to a disaster that befell a Sydney home aquarium during a thunderstorm:

> Lightning shattered a large aquarium on Mr Whiting's verandah drenching his daughters.
> (*Evening News*, 10 December 1879)

Whether keeping your fish outside would have helped, I'm not sure but clearly the appropriately named Mr Whiting thought it did.

There also appears to have been a second wave of aquarium keeping. Aquariums were imported into Tasmania from England in 1876 (*Mercury*, 4 May 1876) and the USA in 1884 (*Daily Telegraph* (Launceston) 24 September 1884) and were for sale at Notman's in Melbourne's Little Collins Street in 1889 (*The Age*, 30 September 1890). There was a prize for the best aquarium "not to exceed three feet by two feet" at the Melbourne Juvenile Exhibition (*The Argus*, 8 March 1879). The South Australian newspaper the *Kapunda Herald* (28 July 1893) republished an article on home aquarium keeping from *The Weekly Scotsman*, while *The Leader* (15 September 1884) contained an article giving rather optimistic instructions as to how to make an aquarium tank. In February and March 1897, the *Australian Town and Country Journal* published a pair of articles on home aquariums that contained more honest advice than many of the advisory pieces published during the first wave of the aquarium mania:

7 Australia: an imperial case study

To manage an aquarium successfully, no matter on how small a scale requires a good deal of care and time, but you will find it time well spent, and the pleasure and knowledge the study of your pets will give you will be an ample return for the time you spend on them.

Notice how the writer ("Prince Hal") gives the fish a place in the domestic sphere and family unit by referring to them as "pets". But I doubt if either article would have been published if the editors had not thought that there was an interested readership for them. So, it appears that, as in England, the aquarium craze never fully went away and had a fin-de-siècle resurgence which has never quite lost momentum.

Aquariums also became part of the urban landscape. The *Sydney Morning Herald* (26 February 1869) alerted readers to:

A curious ornamental adjunct to a shop window, in the shape of an aquarium is now to be seen at the dining room of Mr Holland in Pitt Street.

This was quite a sight and *Empire* (9 February 1869) tells us that it was:

A three feet by two feet ornamental case with a grotto interior, a fountain of dolphins and a Grecian girl and a small flock of miniature birds floating on the water.

It sounds very much like the kind of Cockney aquarium that Hannaford so deplored. But Melbourne was not to be outdone by the cheerful vulgarity of Sydney and by 27 December 1869, the *Illustrated Australian News for Home Readers* announced that the new Royal Arcade in Bourke Street now had an aquarium. Even the little country town of Kilmore had some fish on display with an aquarium at the rear of the local branch of the Bank of Victoria (*North Eastern Ensign*, 16 October 1874).

Tasmania was not to be left behind, with the *Cornwall Chronicle* (3 February 1869) proudly claiming that the aquarium to be seen at Ackermann's Public Baths in Launceston is:

the most complete in its variety of the finny tribe, and the provision for all their wants and comforts ever exhibited.

Nearly ten years later this had grown to a very substantial tank indeed with the *Devon Herald* (14 August 1878) describing "a five thousand fish aquarium where the fish come to the sound of music and feed from their owner's fingers" and another "with two thousand inhabitants supplied with filtered and rarefied water, with floating gardens in the centre driven by an invisible and new motive power" at what was now known as Ackermann's Exhibition. Ackermann, who had started out as a ship's chandler, also sold fish globes and aquarium tanks. His exhibition was still going strong years later with the *Wellington Times* (23 May 1891) telling its readers that they could go there to see:

> a gondola, a gunboat and war canoe decorated from stem to stern with silk flags of all nations, is unsinkable, propelled without an oar, mast or sail, on water teeming with gold and silver fish.

The Launceston *Daily Telegraph* (10 February 1885) mysteriously mentions an £80 payment from the council to a builder for erecting a building "for the marine aquarium" but I can find no reference to what or where this was. Even in late nineteenth-century Launceston, £80 doesn't sound like very much for building anything but a very basic structure. My guess is that it was a shed in the small zoo (known as Button's Menagerie after Alderman Henry Button who established it) that operated as a public amenity in the City Park (and is still residually present there today in the form of an enclosure of Japanese macaques).

Hobart also had its aquariums. The important local philanthropist Morton Allport had an aquarium in the form of a pond in his garden and stocked it by importing fish from England. This was partly an exercise in acclimatisation and partly for the pleasure of having decorative fish. The first shipment was a collection of tench, perch and carp brought in by the redoubtable Captain Harmsworth (who played an important role in bringing English fish, birds and mammals into Tasmania) on the super-fast clipper *Heather Bell*, which plied between London and Australia, always pushing the record time for the voyage. These fish arrived as early as 1858 (*Courier*, 25 February 1858) but

7 Australia: an imperial case study

of 120 fish only six survived and were released successfully into Mr Allport's pond. Another shipment, this time of goldfish, tench and carp, arrived two years later (*Mercury*, 14 September 1860) and, later still, of 100 goldfish shipped in London only five survived the voyage to Hobart because they had been kept in iron tanks that polluted their water (*Mercury*, 22 September 1863). A better solution than this had already been found in the form of the gimble-mounted aquarium designed by William Lloyd and installed in another clipper, the *Lincolnshire*. This was described in the *Mercury* (24 October 1860) and its first shipment contained a wide variety of English freshwater fish including carp, tench, jack, gudgeon, dace and minnows (*Argus*, 13 June 1860). Allport's brother Curzon had a small aquarium in his house in Davey Street (*Mercury*, 11 October 1887). Fish were vulnerable even on relatively short journeys. The minute book of the Zoological and Acclimatisation Society of Victoria for 3 August 1867 records that only half of a shipment of tench sent from Hobart to Melbourne by Morton Allport survived the trip.[4]

But the biggest aquarium in Hobart was that set up by the William Saville-Kent, whom we have met several times before. He was now trying his luck as an ichthyologist in Australia. Saville-Kent had come to Tasmania as Inspector of Fisheries on a three-year contract in 1884. This was on the recommendation of Huxley, who knew him through the British Museum but who also knew Hobart, himself having visited there in 1847 when he was naturalist on board HMS *Rattlesnake*. Saville-Kent's brief was to try to help establish the growing salmon and trout industries, which had been set up in the wake of the first successful transportation of salmon ova from England, to do the same for the oyster industry and generally help with the acclimatisation of fish in Tasmania. Saville-Kent rented a house called "Montezuma" in Napoleon Street in Hobart's Battery Point and there, on the extensive land attached to the house, he set up hatcheries where he experimented with oysters and attempted to establish European lobsters from imported ova.

---

4   On the Acclimatisation Society of Victoria, see Minard, P., *All Things Harmless, Useful and Ornamental* (Chapel Hill: University of North Carolina Press, 2019).

He also built an aquarium which, judging by his sketch of it, was a long low building very much on the model of the original Fish House at Regent's Park. This aquarium was clearly intended as a place for the development of experiments in pisciculture, but it soon established itself as a place that was open to visits from the curious. It had originally been built beside Saville-Kent's first house in Gore Street and was re-erected when he moved to Battery Point. It was 46 feet long and 12 feet wide and contained eight large tanks aerated using a hot air machine which pumped water from a reservoir in the floor which contained 200 gallons. When Saville-Kent's contract expired in 1887, the internal politics of the Fisheries Board meant that his application for another three years was rejected. He went on to serve in Western Australia, Victoria and Queensland and was responsible for important developments in the oyster and pearl fisheries and for some of the earliest significant scientific work on the Great Barrier Reef. Saville-Kent may not have passed muster as an aquarist, at least in Lloyd's mind, but his contribution to the development of marine science and the understanding of the marine environment and maritime ecological systems was immense.[5]

He left the colony soon after, but we know from the inventory of the contents of his house, which were auctioned just prior to his departure (*Mercury*, 29 September 1887), that he was also keeping his own home aquariums. There then arose the problem of what to do with his main aquarium and this was the subject of lively debate in the council. While one councillor described Saville-Kent as an "expensive luxury" who had

---

5   See Harrison, A.J., *Savant of the Australian Seas* (Hobart: Tasmanian Historical Research Association, 1997) for a very detailed account of Saville-Kent's various appointments in Australia and a reproduction of his sketch of his aquarium in Hobart. See also McCalman, I., *The Reef* (New York: Scientific American / Farrar, Strauss & Giroux, 2013) for more detail on Saville-Kent in Queensland and a general account of the discovery of the reef and the development of Australian marine science in the colonial period. As an aside, William Saville-Kent was joined in Hobart by his sister Constance after her release from prison (she served 20 years of a life sentence) who now lived under the name of Emilie Ruth Kaye and travelled with several other of Saville-Kent's siblings who also came to join him. See Summerscale, K., *The Suspicions of Mr Whicher* (London: Bloomsbury, 2008), pp. 283–90.

7 Australia: an imperial case study

achieved little beyond establishing "a toy aquarium" (*Mercury*, 10 June 1887), others were more supportive and pointed out that all the equipment and infrastructure (including that for pumping up sea water) was in place at Battery Point and although it would be too expensive to maintain the aquarium there, it could surely move into the city:

> A great many persons used to visit the aquarium at Battery Point but it was too far out to be generally known, and was not, of course, constructed for exhibition.
> (*Mercury*, 12 July 1887)

Eventually the argument that a municipally owned public aquarium in Hobart would be an asset won the day and it was decided to move it to a room "in the rear of the museum".[6] The council voted £100 for someone to do the work of removing and rebuilding the facility and saved money by letting all the fish go into the Derwent with the somewhat sanguine view that it would be easy enough to catch more fish. This caused some disquiet in the chamber but the Minister of Lands pointed out that it was the infrastructure that mattered, not the fish. "An aquarium was an aquarium even without fish" (*Mercury*, 3 November 1887) he said. "We all know that," groaned the members in response. Perhaps surprisingly. The aquarium was eventually moved but, although some fish were also transferred, the move was a failure and the aquarium was closed by the end of 1888. This incident shows how, even in the 1880s, the idea of an aquarium as a public asset delivering educational and cultural benefits to the citizens still held significant sway in the minds of men who had the public good at heart and the influence and money required to deliver that good where possible. But it is not surprising that the aquarium failed (I cannot find any evidence to show how many people visited it) as once Saville-Kent had gone to seek his

---

6   I am grateful to Ms Janet Carding, the former director of the Tasmanian Museum and Art Gallery, for pointing out to me that, at this time, what we now think of as the front of the museum (the elevation facing Hobart's picturesque harbour) would have been considered the back. It would have made sense to put the aquarium on the side with easy access to the waterfront to facilitate pumping.

fortune in the colonies on the mainland I doubt if there was anyone in Hobart who had the faintest idea of how to maintain healthy tanks and, indeed, the closure of his oyster beds and the removal of the oysters to government-run facilities elsewhere had similar catastrophic results for the oysters.

Although I can find no evidence of aquariums on display in South Australia, there was at least one home aquarium there which, astonishingly, had been successfully transported fully stocked with gold and silver fish from England by a Mr Giles (*South Australian Register*, 2 February 1868). As the figures on fish mortality cited above suggest, this was a quite remarkable feat and the report wisely pointed out:

> The importance of that experiment is less in the value of the fish as Mr Francis has them in great numbers in the Botanic Garden than in the information conveyed as to the proper mode of conveying living fish during a long voyage. The aquarium is common enough in England but it was thought to be too simple an arrangement to protect piscine life in a sea-going vessel.

Two other points of interest here are that the comment about aquariums being common in England suggests that, as late as 1868, they were not in South Australia and that there were fish to be seen in the Adelaide Botanic Gardens. Richard Schomburgk, who succeeded Francis as director, did set up an aquarium. This was an extensive (82 feet by 42 feet) solar-heated water garden. But the comment above suggests that it may also have been stocked with ornamental fish.

Western Australia likewise was behind the game when it came to fish. The *West Australia Times* (8 April 1879) included an aquarium in a long satirical list of things that Perth lacked. But although this may have been a joke, the omission seems to have been remedied the next year when the Alta Pleasure and Picnic Grounds were announced as to open shortly (with no larrikin behaviour allowed). This new park was to include:

> An aquarium for the minor fish and Mollusca of the Swan newly laid out with all kinds of entertainments.
>
> (*The Herald* (Fremantle), 21 February 1880)

7 Australia: an imperial case study

Notice how this proposed aquarium was specifically designed to exhibit local fish, which suggests that although it was in a pleasure garden and thus configured as part of an entertainment there was also, at some level of its mission, the idea that it should be instructive. The pleasure grounds were apparently popular but the debts the proprietor, the ex-convict Thomas Browne, had taken on to develop them meant that they were not a success as a business and had closed by 1882. The account of them in the *Victorian Express* (31 March 1880) makes no mention of the aquarium and it is probable that it was never built.

It is also worth briefly mentioning the Australian Museum in Sydney, especially when it was under the curatorship of Gerard Krefft between 1864 and 1874, as a place where fish could be encountered in quite marvellous ways. This was in the form of taxidermied and photographed specimens of rare and unusual fish that often came from Sydney Harbour. The value of this display was to introduce to the public the idea that the wonders of the deep could be found on their doorstep and that these were specifically Australian wonders. An example was the massive sunfish (1116 kg in weight and 3.4 metres long) that ran ashore in Darling Harbour in 1882 and was presented to the museum where the taxidermy team set to preserving it prior to shipping the skin to the British Museum for stuffing with 25 sacks of straw, a broken chair and a copy of the *Sydney Morning Herald* for 26 January 1883 (which presumably travelled with it from Australia). It was exhibited as part of the Australian display at the International Fisheries Exhibition and then left in the museum (where it still resides). The reason the sunfish didn't come home was that in 1882 three other large sunfish turned up in inshore waters, so the Australian Museum had plenty of its own. One of these was so large that after it was stuffed it was found that it would not fit through the doors, so it was hoisted up the outside of the building and swung in through a large open window on the third floor. The museum was also responsible for the presentation to the scientific world of the Australian lungfish, known as *dala* to the Gubbi Gubbi people of Queensland (who knew all about its capacity to breath out of water) and thus the extension of a specifically Australian Darwinian view of fish evolution.[7] But to understand the importance of the Australian Museum to the encounter with fish, one only has to look at the visitor figures for the Melbourne International Exhibition of 1880 – 1.4 million people (or more than half the population of Australia, which then stood at roughly 2.25 million) could gaze at "three hundred specimens of fish preserved in spirits, with photographs taken from the finest living specimens". The experience must have been a crowded one as this attendance averages out at 6,019 people per day over the course of the exhibition. It is also worth noting that this display was not about fish or marine science in general but about the recognition that Australia had a rich and unique marine environment that was only slowly being understood by the non-Indigenous population.

The next phase of the development of fish displays in Australia once the enthusiasm for home aquariums had taken hold was, predictably, the formation of large public aquariums. Scientists, educators, entrepreneurs, impressarios and philanthropists all followed the development of institutions like the Crystal Palace and Brighton with interest and saw no reason why similar ventures could not succeed in Australia. They seemed less aware of the relative failures of most of the British institutions or, if they were, the optimism that seized investors in the failed aquariums back in the old country seems to have infected them too. And, as in Britain, we will see that different business models, missions and environmental contexts also resulted in different levels of success and failure for the various Australian ventures.

In New South Wales the idea of a public aquarium appears to have been first mooted in December 1874 with the idea of an "aquarium for the people" introduced in the *Sydney Morning Herald* (18 December 1874). The New South Wales Linnean Society also declared itself in favour of an aquarium (*Sydney Morning Herald*, 30 August 1878). The *Maitland Mercury* (29 April 1879) even put a price of £3,000 on this project and claimed that it could be readily stocked with fish from Sydney Harbour. Then, in April 1879, an article in the *Sydney Mail and New South Wales Advertiser* (19 April 1879) proposed a public aquarium in the Sydney Botanical Gardens. "Various worthies" took up the call and wrote to Sir Henry Parkes urging him to establish

---

7   See Finney, V., *Capturing Nature* (Sydney: NewSouth Publishing, 2019) for an informative and wonderfully illustrated account of icthyology (and much else) in the early days of the Australian Museum.

an aquarium either in the Botanical Gardens or, failing that, in the adjacent Domain (*Australian Town and Country Journal*, 3 May 1879). Nothing came of this idea for an aquarium for the colony and it was left to the private sector to begin the process of developing large-scale commercial institutions that would not only enable an encounter with fish but also make money.

This process took several years but on 23 December 1886 a public aquarium opened at Manly. This was a private enterprise led by the property developer Robert Evans (but with a strong municipal interest as Evans was also an alderman). It had a declared educational and scientific mission, although it also had a skating rink as back-up. It was built in only seven weeks and according to the *Sydney Evening News* (1 November 1886) employed 50 labourers. The tanks contained 50,000 gallons of sea water with the seal pond containing another 30,000. The fish were taken from local waters and there were also displays of ferns from Tasmania and shells from New Guinea (*Sydney Morning Herald*, 24 December, 1886). In the first three days, 10,000 people came to visit and it was noted that:

> A curious feature in the exhibition is that the fish appear to be semi-transparent, and some curious optical effects will reward the observer who is careful to watch the effect of the light upon the swimming fish.
>
> (*Sydney Morning Herald*, 28 December 1886)

This comment seems to take us back 30 years to those first people looking at the fish in the Regent's Park Fish House. Even after all this time, there was still novelty and excitement to be had in seeing a fish swimming in a controlled environment and this was now available in Australia. And there was a great deal of difference between seeing the stuffed or preserved specimens in the Australian Museum and these wonderful creatures that coruscated as they swam in and out of the shafts of light that penetrated their tanks.

After a fire in 1887, Robert Evans took over as sole proprietor. There was a scandal in November 1887 when he was prosecuted for animal cruelty for allegedly staging a fight between a seal and a shark. Inspector Webber of the Animal Protection Society saw an advertisement for this and paid his shilling to see it. Two sharks of less than two feet long and with cords round their tails were released into a tank on the side of which sat Bosun the seal, looking on with interest. He plunged in and ate the sharks who did their best to escape. The case came to court. It turned out that this was actually the normal procedure for feeding the seal and that the cords were simply to make it easier to fish out the sharks when their time came. Evans said that the seal wouldn't eat dead sharks, that he had no idea that the advertisement had been placed, that he was sorry that this had been treated as an entertainment and that it wouldn't happen again. The magistrate dismissed the case, but his summing up is instructive:

> "It takes two to make a quarrel and two to make a fight." He added that it was a question whether a shark was an animal. According to the Act [for animal protection and forbidding staged fights and baiting] it was not.
> (*Sydney Morning Herald*, 16 November 1887)

Once again, fish are missing from the circle of moral concern as it is hard to believe that an event involving the feeding of, say, live sheep to wolves, even if that was a normal way of sustaining them, would have attracted the same judgement.

In 1888, Evans attempted to address his worsening financial position by floating the aquarium as a private company but the would-be directors were unconvinced of the long-term solvency of the enterprise and it appears that Evans then used what money he had raised to try to bail out his other business ventures. The aquarium struggled on and even acquired an octopus (*Sydney Morning Herald*, 4 February 1889), so its managers understood the importance of a star animal, especially as they gave it the sensational name of "Jumbo the Diabolical Octopus". The collection generally was of high quality and interest and was clearly part of the development of Manly's sense of itself as a suburb that pulled its weight and as "The Brighton of the South Pacific". It was built in grotto style to emulate the Jenolan Caves and was enthusiastically reviewed in the *Australian Star* (25 March 1889):

7 Australia: an imperial case study

> The collection of fish at the Manly show is unequalled, and it is at any time worthy a visit ... Here one can sit and watch those interesting creatures, the seals, disporting.

But the collection's heyday was also its swansong and by May 1889 the tanks were empty and the building functioned solely as a skating rink with other entertainments, including the hosting of a local institution, The Manly Wild Flower Show, until it closed in 1893. By 1899 the building was demolished:

> The aquarium at Manly which has been so long the resort of visitors and residents is doomed.
> (*The Daily Telegraph*, 6 November 1899)

It was a typical Sydney story: land values on the Manly Corso were simply too high to allow such a site not to be developed for purposes more profitable than an aquarium and skating rink. Another aquarium opened at Manly in late 1901 and so outside our period – this was adjacent to the boardwalk and was swept away in a storm in 1974.

The next aquariums in New South Wales were also in beachfront areas of Sydney, Bondi and Coogee. They were both based unabashedly on an entertainment model where the collection of fish formed part of a more general recreational facility. In this sense they are less interesting than the Manly aquarium even though they were much bigger and lasted longer. Bondi opened in 1887 and continued operating until 1891 (carrying bankruptcy notices from 1890) when it was severely damaged by fire. It reopened later that year as the New Bondi Aquarium. Coogee also opened in 1887, attracted bankruptcy notices in 1895, but managed to struggle on in different forms until 1987.

Bondi was always mainly a place of mass entertainment and most reports of it are for the various shows and exhibitions that it staged such as Blondin the famous tightrope walker or Professor Wyman who could roller-skate on his head. However, the aquarium was properly set up with tanks faced with one-inch-thick plate glass specially commissioned from England. The *Catalogue of Fishes and Other Exhibits at the Royal Aquarium, Bondi* is a substantial volume with the zoological descriptions written by J.D. Ogilby of the Australian

Museum and lists 79 different species in the collection. Entrance was set at 1 shilling for adults, half price for children and, as with the English aquariums, there was a premium day for people who wanted to avoid the crowds – on Wednesday, entrance was half a crown and 1s for children. The aquarium was obviously initially popular. The *Tourist's Pocket Guide to the Blue Mountains, Jenolan Caves, Hawkesbury River and Sights of New South Wales* published by the New South Wales Tourist Bureau in 1888 recommended it and the 10 acres of pleasure grounds that surrounded it. Newspapers reported days in 1889 when several thousand people visited. In April 1889, for example, one day saw over 10,000 visitors and they enjoyed a range of activities:

> Over ten thousand people attended the Bondi Aquarium yesterday, a large number spending all the day there [for] the fish, the fernery, the seals, the camera obscura, merry-go-round, swings, switchback, railways and skating rink.
>
> (*The Daily Telegraph*, 23 April 1889)

From the same report we also learn that there were displays of shooting, skating and performances by trapeze artists. The day was only marred by the poor transport links and the experience (which anyone who has ever lived in Sydney will recognise all too readily) of no co-ordination and dreadful congestion "which appears to completely baffle the tram authorities".

Coogee Aquarium was a very similar enterprise to Bondi and most of the reports and advertisements relating to it are about the various shows it staged: stunt skating, parachute descents, tightrope walking, a herd of donkeys (which doesn't sound very entertaining) and, like Bondi, it struggled to make money even though it enjoyed great popularity as a destination. However, although Bondi might be thought of as an entertainment venue with an aquarium at its heart, Coogee was more like an aquarium that was a small part of an entertainment centre. It is not insignificant that the *Illustrated Guide to the Coogee Aquarium* is only 22 pages long. In fact, the eight tanks containing the fish were set around the auditorium in the way pioneered at Great Yarmouth. The *Daily Telegraph* (24 December 1887) noted the exceptionally clear water (though this was on opening day) and whimsically described the scene:

7 Australia: an imperial case study

the many mystic inhabitants of the deep seem quite at home in their new quarters, and quietly pursue the even tenor of their way without noticing them.

The aquarium was owned by a private consortium. It cost £30,000 to build and incorporated at least some grotto architecture. It was very badly damaged by a gale in June 1890 when it lost much of its roof, and a recently fitted balcony, which had cost £1,600, was ripped off and smashed into the building, damaging it still further, but it was rebuilt – which suggests that three years into the venture the owners were happy with the revenue. It was even noticed as a place to visit after the long voyage from London:

> Wednesday 27th, we went by tram to Coogee and visited the aquarium. It is well worth the visit, though we were told that there is a better one at Bondi.[8]

Given the paucity of information about the fish, this wouldn't be surprising.

Even so, at both Bondi and Coogee the crowds dispersed as the decade wore on – the 1890s simply were not propitious for businesses reliant on discretionary spending in Australia. In addition, although "aquarium" was a magic word that could always stimulate a gleam in a speculative investor's eye, the fact was, as Caroline Ford has well put it, that economic downturn was not a sole factor explanation for business failure:

> No doubt this was partly attributable to the depression of the 1890s, but Sydney's beach crowds were mainly made up of repeat visitors for whom one visit to an aquarium was enough.[9]

The business lessons that were readily available from the failure of aquariums in England were hard to learn.

---

8 Salton, W.M.K., *Notes by the Way: Being the Diary of a Voyage from Australia to London* (Glasgow: Aird & Coghill, 1889), p. 11.
9 Ford, C., *Sydney Beaches: A History* (Sydney: NewSouth Publishing, 2014), p. 22.

Other towns in New South Wales also believed that an aquarium might revive their fortunes. Albury hoped to have one for its local exhibition (*Australian Town and Country Journal*, 28 June 1879) and as late as 1889 there was a plea for an aquarium in Newcastle on the grounds that it would increase visitor numbers to the town (*Newcastle Morning Herald*, 24 April 1889). No doubt the fortunes of Bondi and Coogee were being watched with interest and the dubious logic "if it works in Sydney it will work in Newcastle" was applied. Fortunately for potential investors this plan also came to nothing although it did get as far as a formal discussion in the council chamber (*Newcastle Morning Herald*, 3 July 1889).

If Sydney carried the banner for New South Wales in the form of aquariums as entertainment centres, in Victoria a more sober approach was taken and the planning for an aquarium in Melbourne was based first on an educational and cultural mission and, second, on careful research into the different aquariums in Europe. This may be because the whole approach to animals in Victoria was based on acclimatisation and the zoo was operated by the various incarnations of the Acclimatisation Society of Victoria, which had the very specific mission of developing a new fauna for the colonies and not providing entertainment for the masses. Although acclimatisation plays a role in this book, as we will see in the treatment of Tasmania, there is little evidence that in Melbourne the extensive experiments in the pisciculture of European fish that the society conducted over many years, and especially in the 1860s and 1870s, led to any specific new encounters. For example, whereas Salmon Ponds in Tasmania became a place of resort, the acclimatisation facilities for salmon and trout in Victoria were not developed as attractions and were actually kept secret, with the fish being conveyed at night, to avoid the depredations of poachers.

We first encounter interest in an aquarium in 1874 when an article in *The Herald* (20 April 1874) optimistically stated that an aquarium was "easily constructed from iron and glass" and, this being the case, what was stopping Melbourne from building one? The answer of course is that if a glass and iron aquarium had proved a disastrous design in the foggy chill of London it would be even more disastrous when the January sun beat down in Melbourne. The writer also seems to have little comprehension that aquariums are really about aeration,

7 Australia: an imperial case study

circulation and filtration systems and they are neither easy, nor cheap, to build. *The Leader* (29 August 1874) pointed to the growing popularity of aquariums in both England and America and asked:

When is Victoria to possess a similar institution?

State pride was at stake and a champion was needed. Forward stepped Major Heath. He started the process by going to Europe in 1875 to study aquariums in England and France where he visited Brighton, Manchester, Southport, the Crystal Palace and Paris, and consulted with Frank Buckland. On his return he was able to establish sufficient interest for an aquarium committee to be formed with the aim of funding a large public institution through the sale of 10,000 shares at £1 each. The Heath committee proposed that the new building might be sited next to the corporation baths and that it should incorporate a summer and a winter garden (*Weekly Times*, 11 March 1876). This proposal sparked some local rivalries with a case being made for the aquarium to be sited at St Kilda where there was easy access to sea water and where a deal might be struck with the local railway company to bring people there for the price of 1s, including a bath (*The Telegraph, St. Kilda, Prahran and South Yarra Guardian*, 25 May 1876). *The Argus* (22 March 1876) produced an argument that it should be at Brighton. In fact, Heath and his committee had already written to the city council asking for a grant of land in the city (*The Age*, 22 February 1876).

This request and the argument over location proved academic as by June 1877 Major Heath had to admit defeat and face the fact that the subscription of shares had not been taken up. Even so, he said an aquarium would be popular as 7,000 people "in a few days" had recently gone to see a shark exhibited in Bourke Street (*The Argus*, 29 June 1877). The Melburnian appetite for an encounter with fish was further whetted by the travelling aquarium that accompanied Cooper and Bailey's Circus when it visited in late 1877.[10] The water may have been muddied by a rival project led by a Mr Lee. He had plans for an aquarium with baths and a winter garden with a total project cost

---

10 See Crowley, W.G., *The Australian Tour of Cooper, Bailey & Co.'s Great International Allied Show* (Brisbane: Thorne & Greenwell, 1877).

of £20,000 and wrote asking for a land grant. The relevant minister advised that nothing could be agreed without far more detail and that seems to be the end of Mr Lee's hopes (*The Herald*, 20 June 1877). The interesting thing about this project is firstly the timing – did Lee know that the Heath project had failed to raise the requisite capital and so the field was open? And secondly, the cost – Lee's proposal was twice as expensive as Heath's but perhaps it was more realistic. Heath didn't give up entirely and a year later he was lobbying for space for an aquarium at the forthcoming International Exhibition (*Colac Herald*, 4 October 1878) and, as we shall see, his wish eventually came true.

But it might be asked why it was so difficult to raise money for an aquarium in Melbourne when it was, only a few years later, easy in Sydney. Perhaps the answer comes in a pair of well-argued letters from an anonymous correspondent called Nose U Know published in *The Argus* (7 and 12 May 1876). These set out some sensible and thoughtful objections to the Heath plan. Nose U Know starts by quoting Major Heath's own statement:

> The object of the committee is to give the public an aquarium pure and simple.

But he doubts that this will be sustainable:

> I venture to assert that such an establishment "pure and simple" will soon lose its novelty and fail to attract even Mr Heath's salary as manager.

He further points out that:

> A few fish added to a well-stocked tank will not prove an attractive novelty

that:

> the English aquariums [which Major Heath had visited] are far from "pure and simple" but have all manner of entertainments

7 Australia: an imperial case study

and that:

> a £10,000 show will no more pay in Victoria than it would in London.

A week later Nose U Know returned to the fray. He pointed out that, unlike London, Melbourne had a very wide range of popular entertainments that were free and so the competitive environment for an aquarium that would have to make a profit through an economically viable entry charge was quite hostile.

I doubt whether these letters were, in themselves, enough to sink Heath's project, but they are clearly indicative of the kind of thinking that would-be investors might have been doing and the kind of advice they might have been taking. It is true that, at this time, the evidence of failed aquariums in England would not have been quite so strong as later in the decade but there was enough to cast doubt, especially when the clear fact was that the most successful institutions were by no means just aquariums.

There was another similar plan in 1880 which went so far as to develop a full architectural sketch of a proposed building that would combine an aquarium with a skating rink and with a project plan to raise a capital sum of £6,000. The anonymous projector pointed out that the fish of Australia and its nearby waters were comparatively unknown to world science and that this, in itself, justified a specifically Australian display. He added that:

> The experiment of an aquarium in Melbourne can be made with profit to all concerned
>
> (*The Argus*, 10 January 1880)

This initiative also seems to have gone nowhere.

The notion of an aquarium and an insect house was presented to the board of the Zoological and Acclimatisation Society in February 1883 as part of the director Albert le Souef's plan for developing the zoo after his fact-finding mission to zoos in Europe.[11] This was especially necessary as, since the 1870s, hundreds and sometimes thousands of people were visiting and needed something to keep them distracted and, more importantly, out of mischief while they were on the premises. However, a year later both projects were dropped. In the case of the aquarium this may well have been because the board had learned that a new and extensive facility was about to be established elsewhere in the city.

Victorians finally got their aquarium in 1884 when a wing of the magnificent building constructed for the International Exhibition was repurposed as an aquarium. This was the first large-scale public aquarium in Australia and had several innovative features. Water was carted in from the bay in tanks each containing 4,000 gallons. This water filled three massive underground reservoirs that cumulatively contained 100,000 gallons, which was three times more than was needed to fill the fish tanks and associated pools and ponds. On the face of it, this appears to be the requisite for Lloyd's continuous circulation system but, in fact, this large reserve of water was used to replenish tanks when they became even slightly cloudy. To keep the fish healthy, a powerful compressed air system driven from two cylinders each 30 feet long and three feet in diameter constantly fed streams of air into the tanks. Most of the tanks were made of slate but it was found that sea water eroded this material, so they were then faced with brick. The other tanks were made of glass and iron. The main viewing gallery for the fish was an extensive exercise in grotto architecture. The substantial *Illustrated Handbook to the Aquarium, Picture Salon, Cyclorama, Museum and Technological Collections* listed 46 different kinds of fish as well as seals, penguins, turtles and, innovatively, platypuses. When the aquarium opened it was not fully stocked but this process seems to have been largely completed by the end of 1885.

If the Sydney aquariums were adjuncts to an entertainment complex, the Melbourne Aquarium was located within a rich educational and cultural facility. In spite of this (or maybe because of it), the aquarium, although adequately visited, was never quite the popular success hoped for. A long piece in *The Argus* (22 February 1890) compared the Melbourne facilities with those in Sydney and pointed out that:

---

11   De Courcy, C., *Zoo Story* (Ringwood: Claremont, 1995) is a comprehensive history of Melbourne Zoo.

7 Australia: an imperial case study

In Sydney the first idea of an aquarium appears to have been subordinated to the business of a concert hall and athletics grounds.

Even so, it might be worth considering, "without going so far as the Sydney companies", laying on a few extra entertainments such as afternoon concerts in the grounds as:

> Our aquarium at the present time is unquestionably not appreciated as it should be.

However, the problem might not have been the relentlessly cultural mission but the price: with entrance at 1s for adults and children half price, the cost of a visit was almost prohibitive for a family party of modest means. The writer thought that the increase in footfall resulting from a price reduction to 6d for adults would more than make up the required revenue. These prices compare quite closely to the costs of entry to aquariums in Britain but wages were higher in Victoria and the cost of most common commodities such as milk and tea (but not bread) were lower or comparable, so the paying public would, at least in theory, have more disposable income than their counterparts in, say, Manchester.[12] Perhaps this is why the suggestion to lower prices was not taken up and the aquarium remained a well-visited but not overwhelmingly popular venue until 1953 when it burned down and was not rebuilt.

Stocks for the aquarium came from various places. They were fished systematically from Hastings Bay (although an offer from a fisherman to put his boat exclusively at the aquarium's disposal for a retainer of £300 per annum was turned down). They were also obtained from members of the public who donated curious specimens such as alligators from Queensland, an algae-encrusted crab from

---

12 See James, J.J., *The Emigrant's Guide* (London: International Employment and Emigration Agency, 1882), which is a fascinating compendium of prices and costs not only in Australia but also elsewhere in the empire. See also Franklin, H.M., *A Glance at Australia in 1880* (Melbourne: The Victorian Review Publishing Company, 1881), which is another invaluable source of information about wages and prices.

Warrnambool and the predictable octopus. Seals were also obtained locally but their initial conditions – in a small artificial cave inside the building – were not good, and one visitor described:

> two wretched animals pining and sickening in a puddle beneath a comfortless but most artistic grotto.
>
> (*Hamilton Spectator*, 24 March 1885)

It was clear that seals could not be kept healthily in such conditions, so an outdoor pond was quickly dug for them while a much larger indoor facility (believed to be the largest artificial cave in the world) was also developed. This new environment was a success and by 1890 the correspondent in *The Argus* cited above reported on the extraordinary cleanliness of the water in their ponds and compared it unfavourably with "the green and almost stagnant", slime-infested seal pond at Bondi. But the seals were always well fed, with their fish costing £14 per month (*Mount Alexander Mail*, 21 March 1885). This correspondent also thought that the water used at Melbourne was artificial sea water that performed better than the water pumped straight from the Pacific at Bondi, but this was not the case unless the seal pond, being separate from the main aeration and reservoir system, was replenished with manufactured water.

The English naturalist John Taylor was an early visitor to the Melbourne Aquarium and wrote about it in his *Our Island Continent: A Naturalist's Holiday in Australia*. He was positive in his praise in spite of the faults he saw:

> Unfortunately for myself when I visited the Aquarium, some of the tanks (especially the small table tanks) were out of order. The water was too green ... owing possibly to those tanks having been subjected to too much light. But the larger tanks were in excellent condition, the water was clear and the animals appeared healthy. Pains had been taken to name the animals, and due prominence had been given to the fish, crustacea, zoophytes etc from the Melbourne area.[13]

7 Australia: an imperial case study

Heat and light were always the enemies of the aquarium but what Taylor's account does confirm is the scientific and educational nature of the display with its accurate labelling and an adherence to that part of the institution's mission that was concerned with introducing the fish of Australia to the ichthyological community.

Melbourne Aquarium is a rare, perhaps unique, example of a long-lasting public aquarium based on an educational and cultural mission and although other entertainments did creep in – a cinema was added in the twentieth century – it did fulfil something like Major Heath and his committee's original vision of an aquarium "pure and simple" and, as such, it was remarkably successful. A wonderful illustration of the tanks set in their grotto published in *The Leader* at the very end of our period (31 January 1901) makes the fire of 1953 all the more regrettable as what was lost was not only the collection but also a stunning example of a kind of international architecture specific to a peculiarly Victorian institution. A project for an aquarium in Bendigo mentioned in the *Bendigo Advertiser* (20 November 1877) came to nothing, which, given the distance of that town from the sea and the relative lack of aquarist experience in Victoria at the time, was just as well.

In Queensland an aquarium was developed as part of a bigger building project at the Brisbane suburb of Queensport, which was "to have an aquarium, a township, railways and river steamers" (*Telegraph*, 14 June 1889). This opened on 7 August 1889. It was a two-storey building containing six tanks each 13 by 5 by 4 feet set out in grotto style and incorporating a food court and a 1,400-seat theatre. It stood in 11 acres of landscaped grounds that also included a fernery and a seal pond. The business model, as expressed in the *Queensland Figaro and Punch* (7 September 1889), was that of a general entertainment complex where the aquarium formed part of the attractions on offer in a site of public amusement, and thus it followed what were reasonably successful models in England. It almost immediately ran into problems when the weather intervened:

---

13   Taylor, J.E., *Our Island Continent: A Naturalist's Holiday in Australia* (London: SPCK, 1886), p. 105.

> Unfortunately owing to the recent rains, the water in the fish tanks was muddy, and but little could be seen of the occupants.
>
> (*The Queenslander*, 17 August 1889)

This suggests that the pumping system was drawing water direct from the river or from reservoirs fed by the river. Floods in 1890 killed the seals but, otherwise, the aquarium continued to grow.

It is clear also that the initial popularity of the aquarium, which tried to offer reasonable entrance prices of 2s per adult and 1s per child, which included a return trip on the steamer, soon fell off. This was perhaps as much due to a general economic downturn as shortcomings in the institution itself (although – if my guess about the pumping system is right – I doubt if it was ever possible to keep the tanks very clear). There was also the same problem of having to pay for extras once you were there (as at the Crystal Palace where it also caused irritation). A letter to the *Telegraph* (31 March 1890) made the point well and suggested a solution:

> I find that between one expense and another that a person and his family have to meet with, beyond the actual advertised charge to visit the Aquarium and grounds, the cost seems great.

The writer suggests an all-in charge of 3s (children half-price) to include the boat trip, entrance and all the fairground rides and that there should be no extra charge to listen to the band on the steamer – which sounds odd: one can only assume that if you didn't want to pay to hear a band on a small boat you had to wear earplugs. The aquarium management did eventually offer free entrance to children for several months in 1890 and 1891 and there were reports of significant visitor numbers during this period with days of well over 1,000 people being common. The *Telegraph* (10 November 1891) reported a day of 5,000 visitors "in spite of hard times" and total takings of £11,000. This was at a point when the population of Brisbane stood at about 84,000 so on that day in 1891 roughly one in seven citizens enjoyed an outing to the aquarium. In October 1891 the aquarium offered a free day out with refreshments to Aboriginal children from the Bribie Island Mission. A year later the *Telegraph* (10 November 1892) reported that:

7 Australia: an imperial case study

so satisfactory was the attendance that it seems to be taken as a revival of the early days of the aquarium's existence, when times were good and people generally had the wherewithal to make holiday...

However, this revival was short-lived as in 1893 the weather intervened again and a disastrous flood swept away the contents of the tanks and left the pleasure grounds in ruins. Nobody had the heart to start collecting fish again (and it looks as if it was already known that it would be a difficult venture to make money from anyway) although the aquarium building carried on as a dance hall for a few more years and the steamers were sold. By 1901 the entire site was up for development as building land.

There was also a plan to develop an aquarium linked to the pier in nearby Sandgate, and the mayor and aldermen visited Queensport to see how it was done. The *Telegraph* (3 December 1889) gives us a poignant window into the municipal aspirations of a small Australian town at that time with a savage economic depression just on the horizon:

> I think there is very little doubt it [an aquarium] would be the means of rousing Sandgate from its current lethargy and making it boom along merrily.

By this time, it was well known that it was almost impossible to make money from an aquarium and the people of Sandgate and the local fish were, perhaps, fortunate that nothing came of this scheme. It would be easy to see the demise of the Queensport project as due to an accelerating economic downturn and extreme climate events. However, although these were plainly factors, the underlying problem was, as with most aquariums, that the revenue from the fish, even when added to the extra income from the fairground rides and concerts, simply didn't match the expenditure required. Nevertheless, the citizens of Brisbane didn't go entirely without encountering fish and by 1900 home aquarium keeping was common enough for a Brisbane Aquarium Club publishing its own *Aquarium Bulletin* to be founded and the discussions around the founding of Brisbane Zoo also included the option of an aquarium (*Australian Star*, 16 March 1889).

In South Australia there was talk of establishing an aquarium at Glenelg in Adelaide, but this came to nothing despite some enthusiastic promotion. The *Evening Journal* (13 December 1875) supported the idea but proposed that only an aquarium company would be capable of raising the capital for a project that was certain to be a success as:

> Even a shrimp swimming in a glass of water in a fishmonger's window will attract a crowd of curious onlookers.

The *South Australian Register* (29 May 1876) shows that the project was being actively explored and that William Lloyd had been engaged to advise from London. Interestingly, this article tells us that Lloyd had also been engaged to advise on aquarium projects in Sydney and Melbourne (something I have not seen noted elsewhere). These were also reported on very positively and with the implication that South Australia should not be left behind in the *Adelaide Observer* (8 July 1876). However, by August 1876, the *Express and Telegraph* (29 August 1876) ruefully admitted that the Glenelg project had ended in talk and suggested that Dr Moritz Schomburgk might set up an aquarium in the Adelaide Botanic Garden once the Palm House ("his next 'lion'") is completed. Glenelg had to wait until 1929 until an aquarium was established on the jetty which, alas, was mostly washed away in a great storm in 1948.

Frank Buckland's notions about "economic fish" and fish as food are an integral part of the acclimatisation movement, which is a prominent part of the landscape in colonial Australia. Although it is perhaps a tangential aspect of this study, it is worth pausing on the transportation of live salmon ova to Tasmania and subsequently to other Australian states and to New Zealand. A gap was to be filled. As the *Illustrated London News* (26 January 1867) put it:

> The want of good eatable fish in most of the rivers and streams of Australia and New Zealand is a serious drawback on the comforts and enjoyments of the Colonists.

7 Australia: an imperial case study

Various attempts were made to achieve live export in the 1850s but finally, in 1864, the ship *Norfolk* carrying 100,000 salmon ova and 3,000 trout ova personally selected by Frank Buckland (which must have been a long job and testing on the eyes) and packed in moss-lined cases, the design of which had evolved with each unsuccessful shipment, arrived in Melbourne. A few were retained there but the majority were shipped on to Tasmania where they were transferred to the new Salmon Ponds facility by the River Plenty just north of Hobart. This was now the first commercial fish farm in the southern hemisphere and opened only a year later than the first in the world at Howietoun. All this was due to the efforts of James Youl. He was a rich man who, Magwitch-like, had made a fortune as a grazier in Australia before retiring to England where he became interested in commercial fish breeding to the point of personally using the Crystal Palace Aquarium for experiments and networking with all the major personalities in the economic fish industry, including Frank Buckland and Francis Francis.

The salmon survived. By 1868, salmon were seen swimming in the Derwent and the facility continues to this day as a salmon-breeding establishment.[14] We have already seen the involvement of William Saville-Kent in this initiative and, clearly, the idea of establishing a salmon and trout fishery in Tasmania was considered to be an essential element in the future prosperity of the colony. The introduction of salmon and the facility at Salmon Ponds offered two more ways of encountering fish.

The first was that game fishing was now possible in Tasmania and quickly became established as a respectable leisure pursuit. Trout fishing licences were first offered in 1870. These cost £1 each; 12 people bought them. At a time when the wages of a factory worker might be around 30s a week, this was expensive, especially when the cost of buying suitable fishing equipment is considered.[15]

---

14 Nichols, A., *The Acclimatisation of Salmonidae in the Antipodes, Its History and Results* (London: Samson, Low, Marston, Searle & Rivington, 1882) is a detailed contemporary account of the struggle to establish salmon colonies in Australia. Salmon Ponds still hosts a fascinating museum of trout fishing.
15 See Fahey, C. and Sammartino, A., "Work and Wages at a Melbourne Factory: The Guest Biscuit Works, 1870–1921", *Australian Economic History Review*,

The second was that Salmon Ponds itself became a kind of public aquarium. It is set in attractive grounds, including an interesting collection of trees from around the world, and still makes a good spot for a picnic. In the second half of the nineteenth century, an excursion from Hobart to Salmon Ponds, perhaps on the river steamer *Emu*, was a popular day out and then as now people could look at the hatcheries and also at the various pools and ponds and see various kinds of fully grown salmon and trout swimming, leaping and coming to be fed. As early as 1865, the Tasmanian Temperance Alliance took a day trip to New Norfolk on the *Monarch* (with an "efficient band" on board) and this was extended to allow them plenty of time to see Salmon Ponds (*Mercury*, 31 January 1865). This popularity obviously grew and, by 1867, advertisements for the Bush Hotel in New Norfolk specified that there was:

> A conveyance always on hand for visitors to the Salmon Ponds
>
> (*Mercury*, 5 January 1867)

In 1880, William Senior described Salmon Ponds as:

> One of the institutions of Tasmania, and visitors to Hobart Town are recommended to take the journey to New Norfolk for the sake of the cradle home of the interesting salmon family.[16]

At the end of the century, the Government Printing Office had its annual outing to Salmon Ponds, which were now connected to Hobart by train. The day started with a picnic in the grounds, then a tour of the salmon hatcheries and then, in typical aquarium style, some entertainment – races for the children and a tug o' war between the printers and the pressmen (*Mercury*, 28 January 1900).

The craze for seaweed collecting and conchology was also felt in Australia. There are examples of beautifully constructed seaweed

---

53 (2013), pp. 22–46. *The Argus* (17 August 1870) gave the wages of Melbourne watchmakers as between £4 and £5 per week.

16 Senior, W., *Travel and Trout in the Antipodes* (London: Chatto & Windus, 1880), p. 116.

7 Australia: an imperial case study

albums[17] from the nineteenth century in New South Wales; and, in Tasmania, Mrs Louisa Anne Meredith was, like some of her English and Scottish counterparts, far more than an amateur enthusiast and her marine collections played an important role in the development of marine science as did her books and paintings. The minute books and reports of the Royal Society of Tasmania include many references to shells and shell collections. For example, in May 1871 the society was offered Mr Evans' shell collection for £100.[18] A motion to make Mrs Meredith an honorary member in June 1875 failed in spite of the acknowledgement of her stature as a scientist:

> While full credit was accorded to Mrs Meredith for her very valuable contribution, the feeling of the meeting was that it would be extremely inadvisable to establish the precedent of electing a local resident as an Honorary Member.[19]

Nevertheless, in May 1876, the society thanked her for the presentation of a collection of algae.

This review of the encounter with fish in the Australian colonies demonstrates how a craze in England could be disseminated across the empire with all its phases being reproduced in detail but with a time lag which was surprisingly short given the distances involved and which tells us something about the constant buzz of communication between the main Australian population centres and the old country. Australian aquarium keepers, whether at home or in public commercial ventures, showed all the same optimism and made all the same mistakes as their British counterparts. And, as with their British counterparts, this was often in the face of evidence which suggested that their projects were likely to be overly fraught with risk. The Melbourne Aquarium is a shining exception and perhaps shows that a sustainable public aquarium was really only possible with the municipal support that

---

17 Duggins, M., "Pacific Ocean Flowers: Colonial Seaweed Albums" in Mentz, S. et al., *The Sea and Anglophone Literary Culture* (London: Routledge, 2016), pp. 119–34.
18 Minutes of the Royal Society of Tasmania, 9 May 1871.
19 Minutes of the Royal Society of Tasmania, 8 June 1875.

offered protection against the vagaries of the market and the need to pay dividends. But, even so, the public purse is not unlimited, and the Melbourne experience showed that a reasonably successful business based on culture rather than an entertainment could survive for many years on the basis of only a modestly sized paying public.

Afterword: are the fish still missing?

This chapter is deliberately presented as an Afterword and not a conclusion. The reason for this is, firstly, because I think that the main text contains everything that the reader requires and that a repetition here is superfluous. Secondly, I wanted to shift the perspective and focus, albeit briefly, onto some of the wider issues that may be considered the legacy of the history I have explored in the preceding chapters. In particular, I will deploy material drawing on the legal frameworks that are all that fish have to protect them apart from the caprice of human benevolence. I will also briefly address the nature of the ongoing scientific debate on fish sentience and consciousness.

So far, this book has been largely concerned with the history of aquariums and the related craze for marine science and other maritime animals and materials. It has argued that the development of specific institutions, spaces and technologies in Victorian Britain changed the status of fish and made them visible in entirely new ways. It has had as a thread the way in which fish entered the world of commodities and exchange and the way in which a very typical Victorian craze was structured, how it was determined and who participated in it. Underlying all this though has been a note of concern about the implications of this process for fish and marine creatures. Becoming a thing when you are not a thing (if we allow that sentient beings aren't things) is a dangerous and destructive process. For example, the very nature and quality of your sentience starts to be questioned and in that questioning the logic of reification in capital exchange comes into a difficult and unexpected conjunction with the foundational utilitarianism of the animal rights movement.

A current witticism holds that the first person to mention Hitler loses the argument. At the risk of suffering that fate, I shall suggest that no state has yet passed fish protection laws that equal the *Law on the Slaughter and Holding of Fish and Cold-Blooded Animals* enacted in 1936 in Nazi Germany.[1] A perusal of the present state of animal welfare legislation in Australia, the United Kingdom, the European Union and the United States of America suggests that, in spite of massive advances (relative to a low starting point) in animal welfare and consciousness about animal protection, there is still little address to the growing scientific evidence on pain and distress in fish which would, on the analogy with mammals, lead to greater efforts to define welfare and protection standards designed to minimise suffering and prevent cruelty.

In Australia, where at both federal and state level there are comprehensive animal protection laws (although scant evidence of enforcement), fish are included in general welfare legislation in the Australian Capital Territory, New South Wales, Queensland, Tasmania and Victoria because they are defined as vertebrate animals. In South Australia and Western Australia, they are specifically excluded (that is, they are not seen as animals) and in the Northern Territory they are included only if they are in captivity or dependent on a human for food. But in all these states there are exclusions where fisheries law and regulation override animal welfare codes.

In the United States of America, fish are not covered by the federal *Animal Welfare Act* (first passed in 1966) although some states do have legislation relating to fish. These tend to be around transport and standards of care in fish farms. Legislation devoted to the welfare of fish is often targeted at those areas that are most likely either to optimise yields on fish farms or protect human beings from diseases potentially derived from farmed fish. We see this exemplified in the European Union where the 2020 Guidelines on Fish Welfare are almost entirely directed to the interests of humans by addressing the improvement of

---

1 Sax, B., *Animals in the Third Reich* (Providence: Yogh & Thorn, 2013), p. 104.

conditions in aquacultural contexts.² And this focus is also illustrated in the booklet *Looking Beneath the Surface: Fish Welfare in European Aquaculture* published in 2018 by the Eurogroup for Animals which also has more general interests in fishing and fisheries.³ In 2005 the European Foods Standards Authority commissioned a report that concluded that "all decapods" (that is, crabs, lobsters, prawns etc.) should receive protection. This was on the grounds of the identification of clearly developed awareness and the capacity to feel pain in that class of animals. However, this was not done because of opposition from other scientists. It is interesting to note that of the two options available – a possible risk of causing pain and distress versus a possible risk of not causing pain and distress – the former was preferred. Surely, this is ethically counterintuitive. In 2018, Switzerland, which sensibly remains outside of the European Union, enacted a law protecting crustacea, from, inter alia, being cooked alive. Switzerland remains, so far as I know, the only country to have adopted an approach based on the emerging scientific evidence to this aspect of animal welfare legislation.

In the United Kingdom, fish do have basic rights to protection under the *Animal Welfare Act* (2006) but these are overridden in specific contexts by the *Welfare of Farm Animals (England) Regulation* (2007). However, there is some recognition that understanding of the needs of fish is changing and currently the Department for Environment, Food and Rural Affairs claims that it is working towards higher standards. In addition, the Federation of British Aquatic Societies has issued a substantial and detailed pamphlet, *Your Legal Duty of Care for Keeping Fish under the Animal Welfare Act*, which shows that it does expect its members to be serious about the welfare of the fish in their tanks.⁴

---

2   EU Platform on Animal Welfare Own Initiative Group on Fish, *Guidelines in Water Quality and Handling for the Welfare of Farmed Fish* (2020) (https://feap.info/wp-content/uploads/2020/06/fish-guidelines-june-2020-rev-12.6.2020.pdf)
3   https://www.eurogroupforanimals.org/library/looking-beneath-surface-fish-welfare-european-aquaculture.
4   Federation of British Aquatic Societies, *Your Legal Duty of Care for Keeping Fish under the Animal Welfare Act* (FBAS, 2008) (http://www.fbas.co.uk/FISH%20CARE%20and%20LAW.pdf)

It appears where fish are protected, they are only protected within strict limits and mostly with a view to helping the development of aquaculture. Before going on to consider why this should be, I will first briefly survey the current state of the aquarium industry worldwide.

This book has shown how a boom in aquarium building and general interest in fish swept Britain, Europe and to a lesser extent the USA from the 1850s onwards and had a vibrant and comprehensive manifestation in Australia. It has also shown how the great majority of attempts to keep fish for public display were short-lived and that the fish captured for these institutions had a high mortality rate. Has this changed? Currently, the main business of international aquariums is carried out by large franchises: the Sea World Centres, which are worldwide, and the Sea Life centres, which are found only in the USA and are a branch of the Disney Corporation. In the United Kingdom, many Sea World centres are to be found where once there were aquariums. For example, Brighton Sea World is the latest manifestation of the Brighton Aquarium. These centres are frequently accused of having low welfare standards, high mortality rates among the fish and bogus conservation missions. To explore these issues would require another book so I recommend the interested reader to simply Google "Sea World fish welfare" etc. where he or she can read some of the claims made and the defences put up by the various institutions and make up his or her own mind. The point here is that large global enterprises are now using fish as a commodity for entertainment and display and often contextualising that enterprise with claims about environmental and welfare responsibility. And it is noteworthy that while zoos are now subject to fierce public scrutiny and, in Europe and Australia at least, have tended to clean up their acts where welfare is concerned, the public pressure that has driven that change has not been experienced to anything like the same degree where aquariums and fish are concerned. But, although the nature of the aquarium business may have changed since the nineteenth century, the fundamental commodification of fish on which it relies has proved surprisingly resilient. And one can see that where in the Victorian period entertainment-based aquariums almost always justified their existence by reference to a scientific and educational mission, this justification has been neatly replaced by an appeal to conservation and the environment.

Afterword: are the fish still missing?

Although the terms may have changed, the business, globally, has retained a structural congruence as it has transformed itself in time.

One exception to this depressing story is possibly the case of sea mammals, which have, in parts of this book, been treated as fish because of their ambiguous status in the Victorian mind. There is far more public awareness about the plight of captive dolphins and orcas now and New South Wales has banned dolphin shows in response to this. The highly respected Vancouver Aquarium, which has been a model of best practice in independently verified welfare standards and genuine science, decided to phase out its dolphin and whale programs in advance of the passage of the *Ending the Captivity of Whales and Dolphins Act* into Canadian law in 2019. Most people will have some knowledge of the difficult status of large marine mammals in captivity and of the various international campaigns in favour of whales and dolphins, especially those seeking to disrupt and put an end to commercial whaling and Japanese dolphin hunting (in Japan, sea mammals are generally considered as fish). While all this is progress, it's still worth looking at the reasons behind it, as they will give some sense of why there has been far less advancement in similar campaigns where they relate to fish.

It all comes down to two things. The first is how relatively little is still known about fish. The second is the primacy of sentience, and especially suffering, as a driver for animal welfare concern. While things like the communication systems of dolphins and the migratory routes of whales are increasingly well understood and the subject of large-scale and well-funded study, there is still surprisingly little known about fish and what is known tends to be in support of aquaculture and fishing. For example, it is estimated that between 900 and 2,700 billion wild fish were killed every year between 1999 and 2007 (as opposed to 66 billion chicken, 1.6 billion rabbits,1.5 billion pigs, 1 billion sheep and goats and 0.3 billion cattle).[5] That is a very large number. And note that

---

5   These figures come from fishcount.org.uk/fish-count-estimates-2. As far as I know, this fishcount.org is the only organisation that has attempted to quantify the scale of the global fishing industry. But if anyone doubts these figures on the grounds that they are produced by an organisation with an animal welfare agenda, Peter Singer quotes a very similar figure given by the

the higher estimate is three times the lower estimate. In other words, it is not a very accurate figure compared with the relatively confident numbers attributed to mammals and birds. But even at the lower end it is estimated that about 12 times as many fish are killed for human use as all other food animals put together. This is a quite remarkable figure and throws into sharp relief the relatively low level of activism on fish's behalf (even though in 2018, 79 per cent of people polled in a Eurogroup for Animals survey thought that fish welfare should be improved and a clear majority agreed that fish were sentient, could feel pain and had some form of emotional life) and how little we actually know about what is, apparently, our most common animal-derived food.[6] To add to this, we should include between 460 and 1,100 billion fish killed for oil or meal, 51–167 billion farmed fish killed for food (in 2017 alone) and 250–600 billion farmed shellfish. This doesn't include the trillions of krill hoovered up to support the bogus claims of the fish oil supplement industry and these krill should enter the circle of our concern if only because human intervention in their world deprives whales of their food. And we also cannot accurately compute the number of fish taken illegally. For example, since 1970 it appears that shark and ray numbers have declined by 71 per cent, putting 75 per cent of the various species at risk of extinction. This has almost all been driven by illegal over-fishing to feed Asian markets and as bycatch, especially in the Mediterranean.[7]

Good science on fish pain has been around for a long time and a pioneering work like Victoria Braithwaite's *Do Fish Feel Pain?*, which was first published in 2010 and has since been supported by other studies such as those of Culum Brown and Lynne Sneddon, should have had the same kind of impact as any number of similar studies on the welfare of other animals.[8] Yet fish still stubbornly remain a marginal case for concern and while I doubt if there are any scientific studies now

---

United Nations Food and Aquaculture Organisation ("Fish the Forgotten Victims on Our Plate", *Guardian*, 14 October 2010).

6 Eurogroup for Animals/ CIWF Fish Welfare Survey (comresglobal.com/polls/eurogroup-for-animals-ciwf-fish-welfare-survey/)

7 Pacoureau, N., et al., "Half a Century of Global Decline in Oceanic Sharks and Ray", *Nature*, 589 (2021), pp. 567–71 and Serrano, M. and McNaughton, S., "Disappearing Sharks", *National Geographic* (August 2021), pp. 24–5.

which attempt to justify the vivisection of primates on the grounds of their inferior existential status, there are studies which deny that fish can feel pain.[9]

However, should pain be the threshold that triggers the concern that might lead to protection and a consciousness of welfare needs? Anyone who works in animal studies must respect and admire Peter Singer and acknowledge his foundational role in the formation of a whole range of animal-related issues. However, if there is one thing I would wish for, it would be that I never again see another animal studies publication that, as I did earlier, quotes Bentham on suffering and that the foundational texts of animal rights and welfare thinking had been informed by a proper ethical framework and not the soulless calculus of utilitarianism.[10] Our ability to understand the subjective experience of suffering in another human being can be dangerously imprecise, gets more so when we start to project that experience onto other animals based on mere physiological analogy and becomes even more imprecise when we reach the realms of the cold-blooded and the carapaced. I have suggested above that one of the reasons for the missing fish in animal rights may be the definition of an animal in *Animal Liberation* and now I suggest that the primacy of suffering as a trigger for concern as set out in Bentham is what enables a curious alliance of neglect between proponents of animal welfare and the defenders of the industries and pastimes that depend on killing fish to survive.

If we think we should adopt a different attitude towards fish we do not need a science which demonstrates their potential for suffering. Although it is useful to have it. Nor do we need the wild guesses about whether or not something suffers which populate the utilitarian calculus. What we do need is to consider how, as humans, we use animals (including fish), how our interventions impact on the

---

8   Braithwaite, V., *Do Fish Feel Pain?* (Oxford: Oxford University Press, 2010); Brown, C., Laland, K. and Krause, J., (eds), *Fish Cognition and Behaviour* (Oxford: Blackwell Scientific, 2nd edn, 2011); Sneddon, L., "Pain in Aquatic Animals", *Journal of Experimental Biology*, 218 (2015), pp. 967–76.
9   For example, Key, B., "Why Fish Do Not Feel Pain", *Animal Sentience*, vol 3(1) (https://www.wellbeingintlstudiesrepository.org/animsent/vol1/iss3/1/).
10 Interestingly, in the piece in *The Guardian* cited above, Singer still relies on pain as the determining factor in identifying the victimhood of fish.

environment we share with them and how we might assess the relationship between our benefit and their detriment on the assumption that every living thing would prefer, at some level or other, to be alive rather than dead and to enjoy its environment freely rather than with constraints. We should consider these things irrespective of whether we think the animals can suffer and irrespective of whether we think they have rich, or partial, or no intellectual and emotional capabilities. We should consider solely the right to life and the balance between that and human needs to take or exploit that life.

Once we begin to think in those terms, the enormity of the plight of fish at our hands starts to become clear. What I have attempted to do in this book is to show how the history of Victorian institutions and a Victorian craze created fish as an artefact of industrial civilisation and its specific institutions. The astonishment that the Victorians felt when they saw fish for the first time was so powerful that building machines and creating spaces to reproduce that experience for a mass audience became a significant industry. And, as with most industrial processes, the benefits came with a cost which is still being met.

Bibliography

This bibliography offers a selection of the more important Victorian and Edwardian book sources used for this research and selected secondary sources. Primary sources from contemporary newspapers, magazines and the various council reports and minutes of specific zoos are, for the most part, mostly referenced in the text as they occur and not separately noted here.

Victorian and Edwardian sources

Allom, E.A., *Sea-side Pleasures* (London: Aylott & Jones, 1845).
Allom, E.A., *The Seaweed Collector* (Margate: T. H. Keble, 1841).
Anon., "The Aquarium", *North American Review*, 87 (1858), pp. 143–57.
Anon., *Illustrated Guide to the Coogee Aquarium* (Sydney: Coogee Palace Aquarium Company, 1888).
Anon., *Life beneath the Waves and a Description of Brighton Aquarium* (London: Tinsley Brothers, 1871).
Anon., *London and Its Sights* (London: T. Nelson & Sons, 1859).
Anon., "The Marine Aquarium Madras", *Nature*, 82 (1910), pp. 411–12.
Anon., "Official Guidebook to the Brighton Aquarium [review]", *Nature*, 8 (1873), p. 160.
Anon., "Our Book Shelf", *Nature*, 8 (1873), p. 160.

Anon., "The Vivarium in the London Zoological Gardens", *Illustrated London News*, 15 (1854), pp. 148–55.

Armistead, J.J., *A Short History of the Art of Pisciculture* (Leeds: S. Moxon, 1870).

Atkins, A., *Photographs of British Algae: Cyanotype Impressions* (privately printed for the author, 1843).

Baker, E.A., *Rothesay Royal Aquarium and Its Contents* (Rothesay: The Directors of the Company, 1878).

Bartram, J.E., *The Harvest of the Sea* (London: John Murray, 1873).

Becker, B.H. and Lloyd, W.A., *Official Handbook to the Royal Aquarium and Summer and Winter Garden* (London: The Royal Aquarium, 1876).

Beeton, S.O., *Book of Home Pets* (London: S.O. Beeton, n.d. 1862?).

Bisiker, W., *Across Iceland* (London: Edward Arnold, 1902).

Black, A. and C., *Where Shall We Go?* (Edinburgh: A. & C. Black, 1866).

Blewitt, O., *Panorama of Torquay*, 2nd edn (London: Simpkin and Marshall, Torquay: Cockrem, 1833).

Butler, H., *The Family Aquarium or Aqua Vivarium* (New York: Dick & Fitzgerald, 1858).

Catlow, A., *Drops of Water* (London: Reeve & Benham, 1851).

Catlow, A., *Popular Conchology, or the Shell Cabinet Arranged According to the Modern System* (London: Longman, Brown, Green & Longmans, 1854).

Clarke, L.L., *The Common Seaweeds of the British Coasts and Channel Islands* (London: Frederick Warne & Co., n.d. 1865?).

Croall, A. and Johnstone, W.G., *The Nature-printed British Sea-weeds*, 4 vols (London: Bradbury & Evans, 1859–1860).

Crowley, W.G., *The Australian Tour of Cooper, Bailey & Co.'s Great International Allied Show* (Brisbane: Thorne & Greenwell, 1877).

Cruchley, G.F., *Guide to London: A Handbook for Strangers* (London: G.F. Cruchley, 1862).

Cruikshank, G., "Passing Events or the Tail of the Comet of 1853", *Cruikshank's Magazine*, 1 (1854).

Cunningham, P., *Handbook for London*, 2 vols (London: John Murray, 1849).

Damon, W.E., *Ocean Wonders: A Companion for the Seaside* (New York: D. Appleton, 1879).

Dickens, C., *Dickens's Dictionary of London* (London, 1879).

Dorner, H., *Guide to the New York Aquarium* (New York: D.I. Carson & Co., 1877).

Edwards, A.M., *Life beneath the Waters: or, the Aquarium in America* (New York: H. Baillière, 1858).

Francis, F., *A Book of Angling* (London: Longman, Green & Co., 1876).

Franklin, H.M., *A Glance at Australia in 1880* (Melbourne: The Victorian Review Publishing Company, 1881).

Bibliography

Fraser, R.W., *Ebb and Flow: The Curiosities and Marvels of the Sea-shore* (London: Houlston & Wright, 1860).
Fraser, R.W., *Seaside Divinity* (London: J. Hogg, 1861).
Fraser, R.W., *The Seaside Naturalist* (London: J. Virtue & Co., 1868).
Gatty, M., *British Sea-weeds. Drawn from Professor Harvey's "Phycologia Britannica"* (London: Bell & Daldy, 1863).
Gemmell, P., *A Short History of the Old Green Markets and of the Waverley Market* (Edinburgh: Mould & Todd, 1906).
Gifford, I., *The Marine Botanist* (London: Darton & Co, 1848).
Gordon, E.O., *The Life and Correspondence of William Buckland* (London: John Murray, 1894).
Gosse, E.W., *Father and Son* (London: Heinemann, 1907).
Gosse, E.W., *The Naturalist of the Sea Shore: The Life of Philip Henry Gosse* (London: Heinemann, 1890).
Gosse, P.H., *The Aquarium: An Unveiling of the Wonders of the Deep Sea* (London: John van Voorst, 1854; 2nd edn, 1856).
Gosse, P.H., *A Handbook to the Marine Aquarium* (London: John van Voorst, 1856).
Gosse, P.H., *Natural History: Fishes* (London: SPCK, 1851).
Gosse, P.H., *A Naturalist's Rambles on the Devonshire Coast* (London: John van Voorst, 1853).
Gosse, P.H., *Tenby: A Sea-side Holiday* (London: John van Voorst, 1856).
Gosse, P.H., *A Year at the Shore* (London: Alexander Strahan, 1865).
Handsworth, E.W.H., *Handbook to the Fish House in the Gardens of the Zoological Society of London* (London: Bradbury Evans, 1860).
Hannaford, J., *Seaside and River Rambles in Victoria* (Geelong: Heath & Cordell, 1860).
Harper, J., *Glimpses of Ocean Life* (London: T. Nelson & Sons, 1860).
Harper, J., *The Seaside and Aquarium* (Edinburgh: William P. Nimmo, 1858).
Harvey, W.H., *The Sea-side Book* (London: John van Voorst, 1849).
Hearder, J.N. and Sons, *Guide to Sea Fishing and the Rivers of South Devon* (Plymouth: Hearder & Sons, 1875).
Herdman, W.A., *Port Erin Biological Station: Guide to the Aquarium* (Liverpool: Liverpool Marine Biology Committee, 1906).
Herdman, W.A., "State of the Marine Biology of the Isle of Man", *British Association Handbook* (London: British Association, 1896), pp. 183–9.
Hervey, A.B., *Sea Mosses: A Collector's Guide and an Introduction to the Study of Marine Algae* (Boston: B. Whidden, 1893).
Hibberd, S., *The Book of the Aquarium and Water Cabinet* (London: Groombridge & Sons: 1856).

Hibberd, S., "A Brief History of a Marine Tank", *The Intellectual Observer*, 8 (1866), pp. 259–66.
Hibberd, S., "A Brief History of a River Tank", *The Intellectual Observer*, 7 (1865), pp. 39–45.
Hibberd, S., *Rustic Adornments of Homes of Taste* (London: Groombridge & Sons, 1870).
Hogg, J., *The Microscope: Its History, Construction and Application* (London: Herbert Ingram & Co., 2nd edn, 1856).
Houghton, W., *Sea-side Walks of a Naturalist with His Children* (London: Groombridge & Sons, 1870).
Hughes, W.R., *On the Principles and Management of the Marine Aquarium* (London: John van Voorst, 1875).
Humphreys, W.H., *Ocean Gardens* (London: Simpson, Law & Co., 1857).
Jacobs, R., *Covent Garden: Its Romance and History* (London: Simpkin, Marshall, & Co., 1913).
James, J.J., *The Emigrant's Guide* (London: International Employment and Emigration Agency, 1882).
Jerrold, B., *The Life of George Cruikshank*, 2 vols (London: Chatto & Windus, 1894).
Johnston, G., *An Introduction to Conchology* (London: John van Voorst, 1850).
Kelaart, E.F., "Introductory Report on the Natural History of the Pearl Oyster of Ceylon", *Madras Journal of Literature and Sciences*, 3 (1857), pp. 89–105.
Kingsley, C., *Glaucus: or, The Wonders of the Shore* (London: Macmillan & Co., 1855).
Kingsley, M., *Travels in West Africa* (London: Macmillan, 1897).
Lankester, E., *The Aquavivarium* (London: Robert Hardwicke, 1856).
Lee, H., *The Octopus* (London: Chapman & Hall, 1875).
Lewes, G.H., *Sea-side Studies at Ilfracombe, the Scilly Isles and Jersey* (London: William Blackwood & Sons, 1858).
Lindheimer, O., "Aquarium" in *Gebaude für Sammlungen und Austellung* (Darmstadt: A. Bergstrasser, 1893).
Lloyd, W.A., "Description of the Aquarium of the Royal Aquarium and Summer and Winter Garden", *The Engineer* (October, 1875), pp. 231–4.
Lloyd, W.A., "The Great Public Aquarium at Aston Lower Grounds, Birmingham", *Scientific American Supplement*, no. 189 (1879), pp. 3008–9.
Lloyd, W.A., *A List, with Descriptions, Illustrations and Prices, of Whatever Relates to Aquaria* (London: W. Alford Lloyd, 1858).
Lloyd, W.A., "Notes on the Structure of Aquaria", *The Zoologist*, 11 (1876), pp. 224–32.

Bibliography

Lloyd, W.A., *Official Handbook to the Marine Aquarium of the Crystal Palace Aquarium Company* (London: Crystal Palace Aquarium Company, 1872).
Lovell, M.S., *The Edible Mollusks of Great Britain and Ireland with Recipes for Cooking Them* (London: Reeve & Co., 1867).
Maitland, J.R.G., *History of Howietoun* (Stirling: Howietoun Fishery, 1887).
Martin, J.W., *The Nottingham Style of Float Fishing and Spinning* (London: Sampson, Low, Marston, Searle & Rivington, 1885).
Mather, F., "White Whales in Confinement", *Popular Science Monthly*, 55 (1899), pp. 362–71.
Mayhew, H., *London Labour and the London Poor*, 3 vols (London: Griffin, Bohn & Co., 1861–62).
Meredith, L.A., *Tasmanian Friends and Foes, Feathered, Furred and Finned* (London: M. Ward, 1880).
Mitchell, D., *A Popular Guide to the Gardens of the Zoological Society of London* (London: Printed for the author, 1852).
Mitchell, D., "Zoological Society of London Secretary's Report, February 28, 1854", *The Zoologist*, 12 (1854) pp. 4272–83.
Moggs, E., *Moggs's New Picture of London* (London: E. Moggs, 11th edn, 1848).
Moggs, W., *10,000 Cab Fares from Actual Measurement* (London: W. Moggs, 1859).
New South Wales Tourist Bureau, *Tourist's Pocket Guide to the Blue Mountains, Jenolan Caves, Hawkesbury River and Sights of New South Wales* (Sydney: NSW Government Tourist Bureau, 1888).
Nicols, A., *The Acclimatisation of the Salmonidae at the Antipodes, Its History and Results* (London: Samson, Low, Marston, Searle & Rivington, 1882).
Ogilby, J.D., *Catalogue of Fishes and Other Exhibits at the Royal Aquarium, Bondi* (Sydney: W.M. McLardy, 1887).
Philp, J.M., *The Visitors' Guide to Places Worth Seeing in London* (London: Ward & Lock, 1861).
Pratt, A., *Chapters on the Common Things of the Sea-side* (London: SPCK, 1850).
Rance, C. de, "The Southport Aquarium", *Nature*, 11 (1875), pp. 393–4.
Rideer, J. de and Remoortel, M. van, "'Not Simply Mrs Warren': Eliza Francis Warren (1810–1900) and the *Ladies' Treasury*", *Victorian Periodicals Review*, 44 (2011), pp. 307–26.
Roberts, M., *The Conchologist's Companion* (London: G. & W.B. Whittaker, 1824).
Roberts, M., *A Popular History of the Mollusca* (London: Reeve & Benham, 1851).
Roberts, M., *The Sea-side Companion* (London: Whittaker & Co., 1835).
Routledge, G., *Discoveries and Inventions of the Nineteenth Century* (London: George Routledge & Sons, 1877).
Salton, W.M.K., *Notes by the Way: Being the Diary of a Voyage from Australia to London* (Glasgow: Aird & Coghill, 1889).

Saville-Kent, W., "Aquaria: Their Construction, Management, and Utility", *Journal of the Society of Arts*, 24 (1876), pp. 292–8.

Saville-Kent, W., "The Brighton Aquarium", *Nature*, 8 (1873), pp. 531–3.

Saville-Kent, W., *The Naturalist in Australia* (London: Chapman & Hall, 1897).

Saville-Kent, W., *Official Guide-book to the Brighton Aquarium* (Brighton: The Aquarium Company, 1873).

Saville-Kent, W., *Official Guide-book to the Manchester Aquarium* (Manchester: The Manchester Printing and Stationery Co. Ltd., 1875).

Saville-Kent, W., *Official Handbook to the Royal Aquarium, Westminster* (London: Charles Dickens & Evans, 1877).

Saville-Kent, W., "A Rare Animal at the Manchester Aquarium", *Nature*, 12 (1875), p. 69.

Senior, W., *Travel and Trout in the Antipodes* (London: Chatto & Windus, 1880).

Shenton, F.J.K., *General Guide to the Crystal Palace and Its Gardens* (Sydenham: The Crystal Palace Co., 1886).

Scherren, H., *The Zoological Society of London. A Sketch of its Development and the Story of Its Farm, Museum, Gardens, Menagerie and Library* (London: Cassell & Co., 1905).

Sheppard, H.C., "On the Recent Shells Which Characterise Quebec and Its Environs", *Transactions. Literary and Historical Society of Quebec*, 1 (1829), pp. 188–97.

Sherrard, J.E., *Exhibition Building Melbourne: Illustrated Official Handbook to the Aquarium, Picture Galleries and Museum Collections* (Melbourne: R.S. Brain, 1894).

Small, H.B., "My Aquarium", *The Ottawa Naturalist*, VII (1893), p.2.

Sowerby, G., *A Conchological Manual* (London: H. Bohn, 1852).

Sowerby, G., *A Popular History of the Aquarium* (London: Lovell Reeve, 1857).

Spencer, H., *Education: Intellectual, Moral and Physical* (London: Williams & Norgate, 1861).

Spilling, J., *Molly Miggs's Trip to the Seaside* (London: Jarrold & Sons, n.d., 1870s), p. 68.

Taylor, J.E., *The Aquarium: Its Inhabitants, Structure, and Management* (London: Hardwicke and Bogue, 1876), p.228.

Taylor, J.E., *Half Hours at the Seaside* (London: David Bogue, 1880).

Taylor, J.E., *Our Island Continent: A Naturalist's Holiday in Australia* (London: SPCK, 1886).

Timbs, J., *The Year-Book of Facts in Science and Art* (London: David Bogue, 1853).

Warrington, R., "The Aquatic Plant Case or Parlour Aquarium", *The Garden Companion and Florists' Guide* (1852), pp. 5–7.

Bibliography

Warrington, R., "Notice of Observations on the Adjustment of the Relations between the Animal and Vegetable Kingdoms, by Which the Vital Functions of Both Are Permanently Maintained", *Quarterly Journal of the Chemical Society*, 3 (1851), pp. 52–54.
Warrington, R., "On Artificial Sea Water", *The Annals and Magazine of Natural History*, 2nd Series, 14 (1854), pp. 412–21.
Warrington, R., "On Preserving the Balance Between the Animal and Vegetable Organisms in Sea Water", *The Annals and Magazine of Natural History*, 2nd Series, 12 (1853), pp. 319–24.
Watney, H., "Marine Aquaria", *Hardwicke's Science Gossip*, 7 (1871), pp. 196–8.
Watney, H., "My Crass", *Hardwicke's Science Gossip*, (1871), pp. 12–14.
West, R., "Fresh and Salt Water Aquaria", *Report of the Commissioner of Agriculture for 1864* (Washington: Government Printing Office, 1865), pp. 446–70.
Williams, C.P., *Southern Sunbeams: An Australian Annual for the Field, the River and the Home Circle* (Melbourne: Walker, May & Co., 1879).
Wolf, J., *Zoological Studies Made for the Zoological Society of London from Animals in Their Vivarium, Regent's Park* (London: Henry Graves & Co., 1861–1867).
Woodward, S.P., *A Manual of the Molluscae* (London: J. Weale, 1856).
Wright, E.P., "On the Transportation of Living Fish from South of the Equator to Europe", *The Annals and Magazine of Natural History*, Ser. 4. 2 (1868), pp. 438–41.
Wyatt, M., *Algae Damnonenis, or Dried Specimens of Marine Plants, Principally Collected in Devonshire; Carefully Named According to Dr Hooker's British Flora*, 5 vols (Torquay: Cockrem, for the author, 1834–1840).

Secondary literature

Abberley, W., "Animal Cunning: Deceptive Nature and Truthful Science in Charles Kingsley's Natural Theology", *Victorian Studies*, 58 (2015), pp. 34–56.
Adamowsky, M., *The Mysterious Science of the Sea, 1775–1943* (London: Routledge, 2016).
Allen, D.E., *The Naturalist in Britain* (London: Allen Lane, 1976).
Allen, D.E., "Tastes and Crazes", in N. Jardine, J.A. Secord and E.C. Spary (eds), *Cultures of Natural History* (Cambridge: Cambridge University Press, 1996), pp. 397–407.
Allen, D.E., *The Victorian Fern Craze* (London: Hutchinson, 1969).
Allen, N., Groom, N. and Smith, J. (eds), *Coastal Works: Cultures of the Atlantic Edge* (Oxford: Oxford University Press, 2017).

Amato, S., *Beastly Possessions: Animals in Victorian Consumer Culture* (Toronto: University of Toronto Press, 2015).
Anderson, K., "Coral Jewellery", *Victorian Review*, 34 (2008), pp. 47–52.
Appleby, V., "Ladies with Hammers", *New Scientist* (24 November 1979), pp. 714–18.
Armstrong, I., *Victorian Glassworlds* (Oxford: Oxford University Press, 2008).
Avery, C., *A School of Dolphins* (London: Thames & Hudson, 2009).
Avery, C., Cowie, H.L., Shaw, S. and Wenley, R., *Miss Clara and the Celebrity Beast in Art 1500–1860* (Birmingham: Barber Institute of Fine Arts, 2021).
Bailey, P., *Leisure and Class in Victorian England: Rational Recreation and the Contest for Control 1830–1885* (London: Routledge, 2006).
Baird, G., *Edinburgh Theatres, Cinemas and Circuses, 1820–1963* (Edinburgh: privately printed by the author, 1964).
Barber, L., *The Heyday of Natural History* (London: Jonathan Cape, 1980).
Barker, N., Brodie, A. and Dermott, N., *Margate's Seaside Heritage* (Swindon: English Heritage, 2007).
Barling, S., "See Naples and Dive", *The World of Interiors* (July 2019), pp. 90–9.
Barnard, T., *Nature's Colony* (Singapore: National University of Singapore Press, 2016).
Barrell, J., *The Dark Side of the Landscape* (Cambridge: Cambridge University Press, rev. edn, 1983).
Barrington-Johnson, J., *The Zoo* (London: Robert Hale, 2005).
Bass, M.A., Goldgar, A., Grootenboer, H. and Swan, C., *Conchophilia* (Princeton: Princeton University Press, 2021).
Bechtel, S., *Mr. Hornaday's War* (Boston: Beacon Press, 2012).
Bellanca, M.E., "Recollecting Nature: George Eliot's 'Ilfracombe Journal' and Victorian Women's Natural History Writing", *Modern Language Studies*, 27 (1997), pp. 19–36.
Birkhead, T., *The Wonderful Mr Willughby* (London: Bloomsbury, 2018).
Bloomfield, A.L., *Food and Cooking in Victorian England* (Westport: Praeger Publishers Inc., 2007).
Blunt, W., *The Ark in the Park* (London: Book Club Associates, 1976).
Boase, T., *Mrs Pankhurst's Purple Feather* (London: Aurum Press, 2018).
Boos, A.M., *Goldfish* (London: Reaktion Books, 2019).
Bostridge, M., *Florence Nightingale* (London: Penguin, 2009).
Bosworth, A., "Barnum's Whales, the Showman and the Forging of Modern Animal Captivity", *Perspectives on History, American Historical Association*, 2 April 2018. (www.historians.org).
Braithwaite, V., *Do Fish Feel Pain?* (Oxford: Oxford University Press, 2010).
Briggs, A., *Victorian Things* (London: Batsford, 1988).

Bibliography

Brodie, A. and Whitfield, M., *Blackpool's Seaside Heritage* (Swindon: Historic England, 2015).
Brown, B., "Thing Theory", *Critical Inquiry*, 28 (2001), pp. 1–22.
Brown, C., Laland, K. and Krause, J. (eds), *Fish Cognition and Behaviour* (Oxford: Blackwell Scientific, 2nd edn, 2011).
Browne, J., *Darwin*, 2 vols (London: Jonathan Cape, 1995 and 2002).
Brunner, B., *The Ocean at Home* (London: Reaktion Books, 2011).
Burgess, G.H.O., *The Curious World of Frank Buckland* (London: John Baker, 1967).
Burnett, D.G., *Trying Leviathan* (Princeton: Princeton University Press, 2007).
Cadbury, D., *The Dinosaur Hunters* (London: Fourth Estate, 2000).
Christie, A., "A Taste for Seaweed: William Kilburn's Late Eighteenth-century Designs for Printed Cotton", *Journal of Design History*, 24 (2011), pp. 299–314.
Clark, J.F.M., *Bugs and the Victorians* (New Haven: Yale University Press, 2009).
Coates, P., *Salmon* (London: Reaktion Books, 2006).
Cockayne, E., *Rummage: A History of the Things We Have Reused, Recycled and Refused To Let Go* (London: Profile Books, 2020).
Cohen, D., *Household Gods* (New Haven: Yale University Press, 2009).
Colley, A.C., *Wild Animal Skins in Victorian Britain* (Farnham: Ashgate, 2014).
Collingham, L., *The Hungry Empire* (London: Vintage, 2018).
Coniff, R., "Mad about Sea Shells", *Smithsonian Magazine*, 40 (5) (2009), pp. 44–51.
Conlin, J., *Evolution and the Victorians* (London: A.C. Black, 2014).
Coote, A., Haynes, A.M., Philp, J. and Ville, S., "When Commerce, Science and Leisure Collaborated: The Nineteenth-century Global Boom in Natural History Collections", University of Wollongong, Research online, Faculty of Law, Humanities and Arts, Paper 3221 (https://ro.uow.edu.au/lhapapers/3221).
Corbin, A., *The Lure of the Sea* (London: Penguin, 1995).
Corson, T., *The Secret Life of Lobsters* (New York: Harper Perennial, 2004).
Cowie, H.L., *Exhibiting Animals in Nineteenth-century Britain* (Basingstoke: Palgrave, 2014).
Cowie, H.L., *Victims of Fashion* (Cambridge: Cambridge University Press, 2022).
Crary, J., *Techniques of the Observer* (Boston: MIT Press, 1990).
Crawford, D., *Shark* (London: Reaktion Books, 2008).
Creese, M.R.S., "Fossil Hunters, a Cave Explorer, and a Rock Analyst: Notes on Some Early Women Contributors to Geology", *Geological Society, London, Special Publications*, 281 (January 2007), pp. 39–49.

Creese, M.R.S. and Creese, T.M., *Ladies in the Laboratory III: South African, Australian, New Zealand and Canadian Women in Science: Nineteenth and Early Twentieth Centuries* (Lanham: Scarecrow Press, 2010).

Dance, S.P., "Hugh Cuming (1791-1865), Prince of Collectors", *Journal of the Society for the Bibliography of Natural History*, 9 (1980), pp. 477-501.

Dance, S.P., *Shell Collecting* (London: Faber & Faber, 1966).

Damkaer, D., "John Vaughan Thompson (1779-1847), Pioneer Planktonologist: A Life Renewed", *Journal of Crustacean Biology*, 36 (2016), pp. 256-62.

Davidson, K., "Speculative Viewing: Victorians' Encounters with Coral" in G. Moore and M.J. Smith (eds), *Victorian Environments: Acclimatizing to Change in British Domestic and Colonial Culture* (London: Palgrave Macmillan, 2018), pp. 135-60.

Davis, T.C., "Sex in Public Places: The Zaeo Aquarium Scandal and the Victorian Moral Majority", *Theatre History Studies*, 9 (1990), pp. 1-13.

Day, T., *Sardine* (London: Reaktion Books, 2018).

de Bont, R., *Stations in the Field: A History of Place-based Animal Research 1870-1930* (Chicago: University of Chicago Press, 2015).

De Courcy, C., *Zoo Story* (Ringwood: Claremont, 1995).

De Courcy, C., *Dublin Zoo* (Wilton: The Collins Press, 2009).

Delbourgo, J., *Collecting the World* (London: Penguin, 2018).

Dennis, J., "A History of Captive Birds", *Michigan Quarterly Review*, 53 (3) (2014). (http://hdf.handle.net/2027/spo.act2080.0053.301).

Desmond, A., *Huxley, The Devil's Disciple* (London: Perseus Books, 1997).

Desmond, R., *Sir Joseph Dalton Hooker: Traveller and Plant Collector* (Woodbridge: ACC Art Books, 1999).

Dias, R. and Smith, K., *British Women and Cultural Practices of Empire 1770-1940* (London: Bloomsbury, 2018).

Douglas, M. and Isherwood, B., *The World of Goods* (London: Routledge, 1979).

Duggins, M., "Pacific Ocean Flowers; Colonial Seaweed Albums" in S. Mentz et al., *The Sea and Anglophone Literary Culture* (London: Routledge, 2016), pp. 119-34.

Egerton, F.N., "History of the Ecological Sciences, Part 35, The Beginnings of Marine Biology: Edward Forbes and Philip Gosse", *Bulletin of the Ecological Society of America*, 91 (2010), pp. 176-201.

Ellis, H., *Masculinity and Science in Britain, 1831-1918* (London: Palgrave, 2017).

Elwick, J., *Styles of Reasoning in the British Life Sciences: Shared Assumptions 1820-1858* (London: Routledge, 2015).

Emling, S., *The Fossil Hunter* (New York: St Martin Griffin, 2009).

Endersby, J., *Imperial Nature: Joseph Hooker and the Practice of Victorian Science* (Chicago: University of Chicago Press, 2010).

Bibliography

Estobar, G.F., Corliss, J.O. and Finlay, B.J., "Saville-Kent's String of Pearls", *Protist*, 153 (2002), pp. 413-30.

Fahey, C. and Sammartino, A., "Work and Wages at a Melbourne Factory: The Guest Biscuit Works, 1870-1921", *Australian Economic History Review*, 53 (2013), pp. 22-46.

Feuerstein, A., "Falling in Love with Seaweeds; The Seaside Environments of George Eliot and G. H. Lewes", in L.W. Mazzeno and R.D. Morrison (eds), *Victorian Writers and the Environment* (London: Routledge, 2016).

Finney, V., *Capturing Nature* (Sydney: NewSouth Publishing, 2019).

Flack, A., *The Wild Within* (Charlottesville: University of Virginia Press, 2018).

Flanders, J., *The Victorian House* (London: HarperCollins, 2003).

Ford, C., *Sydney Beaches: A History* (Sydney: NewSouth Publishing, 2014).

Forsberg, L., "Nature's Invisibilia: The Victorian Microscope and the Miniature Fairy", *Victorian Studies*, 57 (2015), pp. 638-66.

Fort, T., *The Book of Eels* (London: HarperCollins, 2003).

Freeberg, E., *A Traitor to His Species* (New York: Basic Books, 2020).

Fyfe, A., "Natural History and the Victorian Tourist: From Landscapes to Rockpools", in D.N. Livingstone and C.W.J. Withers (eds), *Geographies of Nineteenth-century Science* (Chicago: University of Chicago Press, 2011), pp. 371-98.

Garascia, A., "'Impressions of Plants Themselves': Materializing Eco-archival Practices with Anna Atkins's *Photographs of British Algae*", *Victorian Literature and Culture*, 47 (2019), pp. 267-303.

Gates, B., "Introduction: Why Victorian Natural History?", *Victorian Literature and Culture*, 35 (2007), pp. 539-49.

Gates, B., *Kindred Nature* (Chicago: Chicago University Press, 1998).

Girling, R., *The Man Who Ate the Zoo* (London: Chattto & Windus, 2016).

Godfrey-Smith, P., *Other Minds: The Octopus and the Evolution of Intelligent Life* (London: William Collins, 2018).

Gooday, G., "'Nature' in the Laboratory: Domestication and Discipline with the Microscope in Victorian Life Science", *British Journal for the History of Science*, 24 (1991), pp. 307-41.

Granata, S., "At Once Pet, Ornament and Subject for Dissection: The Unstable Status of Marine Animals in Victorian Aquaria", *Cahiers Victoriens et Édouardiens* (en ligne), 88 (2018).

Granata, S., "The Dark Side of the Tank", in H. Kingstone and K. Lister (eds), *Paraphernalia! Victorian Objects* (London: Routledge, 2018), pp. 48-58.

Granata, S., "Let Us Hasten to the Beach: Victorian Tourism and Seaside Collecting", *Lit: Literature, Interpretation, Theory*, 27 (2016), pp. 91-110.

Granata, S., *The Victorian Aquarium, Literary Discussions on Nature, Culture and Science* (Manchester: Manchester University Press, 2021).
Gray, A., *The Greedy Queen* (London: Profile Books, 2018).
Greenwood, P.H., "The Zoological Society and Ichthyology, 1826–1930", *Symposium of the Zoological Society of London*, 40 (1976), pp. 85–104.
Grigson, C., *Menagerie* (Oxford: Oxford University Press, 2016).
Groeben, C., "The Stazione Zoological Anton Dohrn as a place for the circulation of scientific ideas: Vision and Management", in K.L. Anderson and C. Thierry (eds), *Information for Responsible Fisheries: Libraries as Mediators*, Proceedings of the 31st Conference (2006), pp. 291–9.
Grootenboer, H., "Thinking with Shells in Petronella Oortman's Dollhouse", in M.A. Bass, A. Goldgar, H. Grootenboer and C. Swan, *Conchophilia* (Princeton: Princeton University Press, 2021), pp. 103–26.
Guillery, P., *The Buildings of London Zoo* (London: Royal Commission on the Historic Monuments of England, 1993).
Haddad, A., "Nature for Ladies: The Victorian Art of Flower and Seaweed Pressing", *Merchant's House Museum* (https://merchantshouse.org/blog/seaweed-pressing, 21 August 2017).
Hadjiaxfendi, K. and Plunkett, J., "Science at the Seaside: Pleasure Hunts in Victorian Devon", in N. Allen, N. Groom and J. Smith (eds), *Coastal Works: Cultures of the Atlantic Edge* (Oxford: Oxford University Press, 2017), pp. 181–204.
Hamera, J., *Parlor Ponds, the Cultural Work of the American Home Aquarium 1850–1970* (Ann Arbor: University of Michigan Press, 2011).
Hamlin, C., "Robert Warrington and the Moral Economy of the Aquarium", *Journal of the History of Biology*, 19 (1986), pp. 131–53.
Harrison, A.J., *Savant of the Australian Seas* (Hobart: Tasmanian Historical Research Association, 1997).
Harter, U., *Aquaria in Kunst, Literatur und Wissenschaft* (Heidelberg: Kehrer Verlag, 2014).
Hausman, K. and Machemer, H., "The Microcosm under the Microscope: A Passion of Amateurs and Experts", *Denisia*, 41 (2018), pp. 1–46, p. 12.
Hibbert, C., *The Great Mutiny* (Harmondsworth: Allen Lane, 1978).
Hilton, T., *The Pre-Raphaelites* (London: Thames & Hudson, 1970).
Hoage, R.J. and Deiss, W.A. (eds), *New Worlds, New Animals* (Baltimore: Johns Hopkins University Press, 1996).
Hoare, P., *Albert and the Whale* (London: Pegasus Books, 2021).
Hoare, P., *Leviathan, or, The Whale* (London: Fourth Estate, 2008).
Hodkinson, I.D. and Steward, A., *The Three-legged Society* (Lancaster: Centre for North-Western Regional Studies, 2012).

Bibliography

Holmes, J., *The Pre-Raphaelites and Science* (New Haven: Yale University Press, 2018).
Holway, T., *The Flower of Empire* (Oxford: Oxford University Press, 2013).
Hoskins, I., *Coast* (Sydney: NewSouth Publishing, 2013).
Howe, M., *Old Rhyll from the 1850s to 1910* (Rhyll: Gwasg Helydd, 2000).
Hughes, K., *The Victorians Undone* (London: Fourth Estate, 2018).
Hunt, S.E., "Free, Bold, Joyous: The Love of Seaweed in Margaret Gatty and Other Mid-Victorian Writers", *Environment and History*, 11 (2005), pp. 5–34.
Ingemark, "The Octopus in the Sewers: An Ancient Legend Analogue", *Journal of Folklore Research*, 45 (2008), pp. 145–70.
Ingle, R., "Who was William Alford Lloyd?", *Biologist*, 60 (1 December 2013), pp. 24–7.
Ingleby, M., "'Human Language Can Make a Shift', Late Victorian Tentacular Cities and the Genealogy of 'sprawl'" in W. Parkins (ed.), *Victorian Sustainablity in Literature and Culture* (London: Routledge, 2017), pp. 146–64.
Ito, T., *London Zoo and the Victorians 1829–1859* (Woodbridge: The Royal Historical Society, 2014).
Jackson, C.E., *Menageries in Britain 1100–2000* (London: The Ray Society, 2014).
Jackson, L., *Palaces of Pleasure* (New Haven: Yale University Press, 2019).
Jackson, P.N.W., "Robert Ball (1802–1857): Naturalist", *The Irish Naturalist's Journal*, 30 (2009), pp. 15–18.
Jardine, N., Secord, J.A. and Spary, E.C., *Cultures of Natural History* (Cambridge: Cambridge University Press, 1996).
Johnson, K.W., *The Feather Thief* (London: Hutchinson, 2018).
Keeling, C.H., "Zoological Gardens of Great Britain", in V.N. Kisling, ed., *Zoo and Aquarium History* (Boca Raton: CRC Press, 2001), pp. 49–74.
Kennedy, M., "Discriminating the 'Minuter Beauties of Nature', Botany as Natural Theology in a Victorian Medical School", in L. Karpenko and S. Claggett (eds), *Strange Science: Investigating the Limits of Knowledge in the Victorian Age* (Ann Arbor: University of Michigan Press, 2017), pp. 40–61.
Keogh, L., *The Wardian Case* (Kew: Kew Publishing, 2020).
Key, B., "Why Fish Do Not Feel Pain", *Animal Sentience*, vol 3(1), (https://www.wellbeingintlstudiesrepository.org/animsent/vol1/iss3/1/)
King, A.M., *Bloom: The Botanical Vernacular in the English Novel 1770–1900* (Oxford: Oxford University Press, 2003).
King, A.M., *The Divine in the Commonplace: Reverent Natural History and the Novel in Britain* (Cambridge: Cambridge University Press, 2019).
King, A.M., "George Eliot and Science" in G. Levine and N. Henry (eds), *The Cambridge Companion to George Eliot* (Cambridge: Cambridge University Press, 2019), pp. 175–94.

King, A.M., "Reorienting the Scientific Frontier: Victorian Tide Pools and Literary Realism", *Victorian Studies*, 47 (2005), pp. 153–63.
King, A.M., "Tide Pools", *Victorian Review*, 36 (2010), pp. 40–5.
King, R.J., *Lobster* (London: Reaktion Books, 2011).
Kingstone, H. and Lister, K. (eds), *Paraphernalia! Victorian Objects* (London: Routledge, 2018).
Kisling, V.N., ed., *Zoo and Aquarium History* (Boca Raton: CRC Press, 2001).
Klee, A., *The Toy Fish: A History of the Aquarium Movement in America* (Pascoag R. I.: Finley Aquatic Books, 2003).
Klee, A.J., "Who Invented the Aquarium?" *Aquarium Hobby Historical Society Files*, 17 November 2012.
Kofoid, C.A., *The Biological Stations of Europe* (Washington: Government Printing Office, 1910).
Kraus, A. and Winkler, M. (eds), *Weltmeere: Wissen und Wahrmehrung im Langen 19 Jahrhundert* (Göttingen: Vandenhoeck & Ruprecht, 2014).
Kurlansky, M., *The Big Oyster* (New York: Random House, 2007).
Kurlansky, M., *Cod* (London: Vintage, 1999).
Lambrechts, W., "Brussels Zoo: A Mirror of Nineteenth-century Modes of Thought on the City, Science and Entertainment", *Brussels Studies* (2014). (https://journals.openedition.org/Brussels/1223#tictoZnl).
Latham, G., et al., *The Palace and Park* (London: Good Press, 2019).
Le Gall, G., "Dioramas Aquatiques: Théophile Gautier Visite l'Aquarium du Jardin d'Acclimatation", *Culture et Musées*, 32 (2018), pp. 81–106.
Lever, C., *They Dined on Eland* (London: Quiller Press, 1992).
Lightman, B., *Victorian Popularizers of Science* (Chicago: University of Chicago Press, 2007).
Linzey, A. and Linzey, C. (eds), *The Global Guide to Animal Protection* (Urbana: University of Illinois Press, 2013).
Livingstone, D.N. and Withers, C.W.J. (eds), *Geographies of Nineteenth-century Science* (Chicago: University of Chicago Press, 2011).
Locker, A., *Freshwater Fish in England* (Oxford: Oxbow Books, 2018).
Locker, A., "The Social History of Coarse Angling in England AD 1750–1950", *Anthropozoologica*, 49 (2014), pp. 99–107.
Loisel, G.A.A., *Histoire des Ménageries*, 3 vols (Paris: Octave Doin et Fils, 1912).
Lorenzi, C., "L'engouement pour l'Aquarium en France (1855–1870)", *Sociétés et Représentations*, 28 (2009), pp. 253–71.
MacDonald, R. and McMillan, N., "James Dowsett Rose Cleland (Cleland: A Forgotten Irish Naturalist", *The Irish Naturalist's Journal*, 13 (1959), pp. 70–72.
Maddox, B., *Reading the Rocks* (London: Bloomsbury, 2017).

Bibliography

Martinez, A., "A Souvenir of Undersea Landscapes: Underwater Photography and the Limits of Photographic Visibility", *História Ciéncias Saûde-Manguinhos*, 21 (2014), pp. 1024–41.

Mauriès, P., *Coquillages et Rocailles* (Paris: Thames & Hudson, 1994).

Mazzeno, L.W. and Morrison, R.D., *Victorian Writers and the Environment* (London: Routledge, 2016).

McCalman, I., *The Reef* (New York: Scientific American / Farrar, Strauss & Giroux, 2013).

McDowell, R.M., *Gamekeepers for the Nation* (Christchurch: Canterbury University Press, 1994).

McKie, E., *Thomas Hall of City Road: The Family in a Museum* (London: Create Open Space Independent Books, 2013).

McMurray, D., *A Rod of Her Own: Women and Angling in Victorian North America*, M.A. thesis, University of Calgary, 2007.

Mehos, D., *Science and Culture for Members Only: The Amsterdam Zoo Artis in the Nineteenth Century* (Amsterdam: Amsterdam University Press, 2006).

Mentz, S. et al., *The Sea and Anglophone Literary Culture* (London: Routledge, 2016).

Merill, L., *The Romance of Victorian Natural History* (Oxford: Oxford University Press, 1989).

Minard, P., *All Things Harmless, Useful and Ornamental* (Chapel Hill: University of North Carolina Press, 2019).

Moine, F., *Women Poets in the Victorian Era: Cultural Practices and Nature Poetry* (London: Routledge, 2016).

Montgomery, S., *The Soul of an Octopus* (New York: Simon & Schuster, 2015).

Moore, G. and Smith, M.J. (eds), *Victorian Environments: Acclimatizing to Change in British Domestic and Colonial Culture* (London: Palgrave Macmillan, 2018).

Moore, P., "Seaside Natural History and Divinity: A Science-inclined Scottish Cleric's Avoidance of Evolution (1860–1868)", *Archives of Natural History*, 40 (2013), pp. 84–93.

Moss, S., *Birds Britannia* (London: HarperCollins Publishers, 2011).

Munro, J.M., *The Royal Aquarium: Failure of a Victorian Compromise* (Beirut: American University of Beirut, 1971).

Murphy, J.B. and McCloud, K., "The Evolution of Keeping Captive Amphibians and Reptiles", *Herpetological Review*, 41 (2010), pp. 134–42.

Murphy, J.B., *Herpetological History of the Zoo and Aquarium World* (Malabar: Krieger Publishing Company, 2007).

Nicolson, A., *The Sea Is Not Made of Water* (London: HarperCollins, 2021).

Novak, J. and Patoka, J., "Modern Ornamental Aquaculture in Europe: Early History of Fresh Water Fish Imports", *Reviews in Aquaculture*, 12 (2020), pp. 1–19.

Nyhart, L.K., *Modern Nature: The Rise of the Biological Perspective in Germany* (Chicago: University of Chicago Press, 2009).

O'Connor, K., *Seaweed, A Global History* (London: Reaktion Books, 2017).

Orr, M., "Fish with a Different Angle: *The Fresh-Water Fishes of Great Britain* by Mrs Sarah Bowdich (1791–1856)", *Annals of Science*, 71 (2014), pp. 206–40.

Orr, M., "Women Peers in the Scientific Realm: Sarah Bowdich (Lee)'s Expert Collaborations with Georges Cuvier, 1825–1833", *Notes and Records of the Royal Society*, 69 (2015), pp. 37–51.

Owen, J., *Trout* (London: Reaktion Books, 2012).

Pacoureau, N., et al., "Half a Century of Decline in Oceanic Sharks and Rays", *Nature*, 589 (2012), pp. 567–71.

Parkins, W. (ed.), *Victorian Sustainability in Literature and Culture* (London: Routledge, 2017).

Pearson, L., *The People's Palaces: The Story of the Seaside Pleasure Buildings of 1870–1914* (Northampton: Quotes Ltd, 1991).

Pepperel, J., *Fishing for the Past* (Dural: Rosenberg, 2018).

Picard, L., *Victorian London* (London: Weidenfeld & Nicolson, 2005).

Pierce, P., *Jurassic Mary* (Stroud: The History Press, 2006).

Plaiser, H., Bryant, J.A., Irvine, L.M., Jones, M. and Spencer Jones, M.E., "The Life and Work of Margaret Gatty (1809–1873), with Particular Reference to her Seaweed Collections", *Archives of Natural History*, 43 (2016), pp. 336–50.

Pooley, S., "Parenthood, Child-rearing and Fertility in England, 1850–1914", *History of the Family*, 18 (2013), pp. 83–106.

Priebe, J.M.S., *Conchyliologie to Conchyliomanie: The Cabinet of François Boucher*, PhD thesis, University of Sydney, 2011.

Purdon, J., "At the Aq", *Junket*, 2 (9 January 2012), thejunket.org/2012/01/issue-two/at-the-aq.

Rauch, A., *Dolphin* (London: Reaktion Books, 2013).

Ray, D. and Roberts, M., *Swimming Communities in Victorian England* (London: Palgrave, 2019).

Rehbock, P.F., "The Victorian Aquarium in Ecological and Social Perspective", in M. Sears and D. Merriman (eds), *Oceanography: The Past* (New York: Springer), pp. 522–39.

Reidy, M.S. and Rozwadowski, H.M., "The Spaces in Between: Science, Ocean, Empire", *Isis*, 105 (2014), pp. 338–51.

Reiss, C., "Gateway, Instrument, Environment: The Aquarium as a Hybrid Space between Animal Fancying and Experimental Zoology", *International Journal*

Bibliography

of History and Ethics of Natural Sciences, Technology and Medicine, 20 (2012), pp. 309–36.

Ricke, H., "Der Vogel im Goldfischglas" (The bird in the goldfish bowl), *Journal of Glass Studies*, 59 (2017), pp. 261–84.

Ridder, J. and Remoortel, M. van, "Not 'Simply Mrs. Warren': Eliza Warren Frances (1810-1900) and the *Ladies' Treasury*", *Victorian Periodicals Review*, 44 (2011), pp. 307–26.

Roman, J., *Whale* (London: Reaktion Books, 2005).

Rudwick, M.J.S., *The Great Devonian Controversy* (Chicago: Chicago University Press, 1985).

Rudwick, M.J.S., *Worlds before Adam: The Reconstruction of Geohistory in the Age of Reform* (Chicago: Chicago University Press, 2008).

Rupke, N., *Richard Owen* (Chicago: Chicago University Press, 2009).

Ryan, J., *The Forgotten Aquariums of Boston* (Pascoag: Finley Aquatic Books, 2002).

Saunders, B., *Discovery of Australia's Fishes* (Collingwood: CSIRO Publishing, 2012).

Sax, B., *Animals in the Third Reich* (Providence: Yogh & Thorn, 2013).

Saxon, A.H., "P.T. Barnum and the American Museum", *Wilson Quarterly*, 13 (1989), pp. 130–9.

Schaaf, L.J., *Sun Gardens, Cyanotypes by Anna Atkins* (New York: New York Public Library, 2018).

Schweid, R., *Octopus* (London: Reaktion Books, 2014).

Sears, M. and Merriman, D. (eds), *Oceanography: The Past* (New York: Springer, 1980).

Senior, E., "'Glimpses of the Wonderful': The Jamaican Origins of the Aquarium", *Global Currents*, 19 (2022), pp. 128–52.

Serrano, M. and McNaughton, S., "Disappearing Sharks", *National Geographic* (August 2021), pp. 24–5.

Sherson, E., *London's Lost Theatres of the Nineteenth Century* (London: John Lane, 1925).

Shetterly, S.H., *The Seaweed Chronicles* (New York: Algonquin Books, 2018).

Shick, J.M., *Where Corals Lie* (London: Reaktion Books, 2018).

Shteir, A.B., "Gender and 'Modern Botany' in Victorian England", *Osiris*, 12 (1997), pp. 29–38.

Simons, J., *Obaysch: A Hippopotamus in Victorian London* (Sydney: Sydney University Press, 2018).

Simons, J., *Rossetti's Wombat* (London: Middlesex University Press, 2008).

Simons, J., *The Tiger That Swallowed the Boy: Exotic Animals in Victorian England* (Faringdon: Libri Publishing, 2012).

Singer, P., *Animal Liberation* (New York: Avon Books, 1975).
Singer, P., "Fish the Forgotten Victims on Our Plate", *The Guardian*, 14 October 2010.
Smith, J., *Charles Darwin and Victorian Visual Culture* (Cambridge: Cambridge University Press, 2006).
Smith, J., "Eden under Water: The Visual Natural Theology of Philip Gosse's Aquarium Books", paper presented at *Nineteenth-Century Religion and the Fragmentation of Culture in Europe and America* (Lancaster University, 1997) (www.personal.umd.umich.edu/~jonsmith/PHGINCS.html)
Sneddon, L., "Pain in Aquatic Animals", *Journal of Experimental Biology*, 218 (2015), pp. 967–76.
Sowerby, G., *Sowerby's Book of Shells* (New York: Crescent Books, 1990).
Stamp, G., *Lost Victorian Britain* (London: Aurum, 2013).
Stevenson, B., "Joseph Sinel, 1844–1929, James Hornell, 1865–1949", microscopist.net
Stocks, C. and Lewin, A., *The Book of Pebbles* (Norwich: Random Spectacular, 2018).
Stott, R., *Darwin and the Barnacle* (London: Faber & Faber, 2005).
Stott, R., *Oyster* (London: Reaktion Books, 2004).
Stott, R., *Theatres of Glass* (London: Short Books, 2003).
Stott, R., "Through a Glass Darkly. Aquarium Colonies and Nineteenth-century Narratives of Marine Monstrosity", *Gothic Studies*, 2 (2018), pp. 305–27.
Strehlow, H., "Zoos and Aquariums of Berlin", in R.J. Hoage and W.A. Deiss (eds), *New Worlds, New Animals* (Baltimore: Johns Hopkins University Press, 1996), pp. 63–72.
Summerscale, K., *The Suspicions of Mr Whicher* (London: Bloomsbury, 2008).
Szczygieska, M., "Elephant Empire: Zoos and Colonial Encounters in Eastern Europe", *Cultural Studies*, 34 (2020), pp. 789–810.
Teltscher, K., *Palace of Palms: Tropical Dreams and the Making of Kew* (London: Picador, 2020).
Thompson, R., "Women in the Fishing: The Roots of Power between the Sexes", *Comparative Studies in Society and History*, 27 (1985), pp. 3–32.
Thwaite, A., *Glimpses of the Wonderful* (London: Faber & Faber, 2002).
Travis, J.F., *The Rise of the Devon Seaside Resorts* (Exeter: University of Exeter Press, 1993).
Trentmann, F., *Empire of Things* (London: Penguin, 2017).
Valen, D., "On the Horticultural Origins of Victorian Glasshouse Culture", *Journal of the Society of Architectural Historians*, 75 (2016), pp. 403–23.
Vasile, R.S., *William Stimpson and the Golden Age of American Natural History* (Ithaca: Cornell University Press, 2018).

Bibliography

Velten, H., *Beastly London* (London: Reaktion Books, 2013).
Vennen, M., "Echte Forscher und wahre Liebhaber, Der Blick ins Meer durch das Aquarium im 19 Jahrhundert", in A. Kraus and M. Winkler (Eds), *Weltmeere: Wissen und Wahrmehrung im Langen 19 Jahrhundert* (Göttingen: Vandenhoeck & Ruprecht, 2014), pp. 84–102.
Vennen, M., *Das Aquarium* (Göttingen: Walstein Verlag, 2018).
Vevers, G., *London's Zoo* (London: Bodley Head, 1976).
Wallen, M., *Squid* (London: Reaktion Books, 2021).
Warer, D., "Seaside Microscopy: A Favorite Victorian Hobby", 7 August 2015. (americanhistory.sci.ed)
Warner, D., "William Stimpson and the Smithsonian's First Aquarium", *Smithsonian Institution Archives*, 24 July 2015. (siarchives.si.edu)
Whittingham, S., *Fern Fever* (London: Frances Lincoln, 2009).
Wijgerde, T., "Victorian Pioneers of the Marine Aquarium", *Advanced Aquarist* (February 2016) (reefs.com/magazine/Victorian-pioneers-of-the-marine-aquarium/)
Wilkinson, A., "The Preternatural Gardener: The Life of Shirley James Hibberd, 1825–1890", *Garden History*, 26 (1998), pp. 153–75.
Wilkinson, A., *Shirley Hibberd: The Father of Amateur Gardening* (Bromsgrove: Cortex Design, 2012).
Yallop, J., *Magpies, Squirrels and Thieves* (London: Atlantic Books, 2011).
Yonge, C.M., *British Marine Life* (London: William Collins, 1944).
Zon, B., *Evolution and Victorian Musical Culture* (Cambridge: Cambridge University Press, 2017).

Index

Aelian 2
Agardh, Jacob Georg 116
Albert, Prince of Monaco 147
Allen, David 118
Allom, Elizabeth 99
Allport, Morton 238–239
Amato, Sarah 22
Anderson, Katherine 132–133
Anning, Mary 115, 139
aquariums
  Adelaide 242, 260
  American Museum 17, 162
  Artis Zoo (Amsterdam) 163
  Aston (Birmingham) 151, 198, 228
  Berlin 89, 161, 163, 167
  Blackpool 198, 199, 210
  Bondi 247–250, 256
  Boston 162, 163
  Brighton 1, 63, 151, 161, 171, 184–195, 199, 203, 215, 268
  Brisbane (Botanical Gardens) 234
  Brisbane (Queensport) 257, 259
  Brussels 167
  Coogee 247–250
  Crystal Palace 20, 54, 151, 161, 169–184, 188, 196, 199, 208, 258, 261
  Derby 198
  Douglas 226
  Dublin (Aquatic Vivarium) 64–69, 167, 173
  Edinburgh (Waverley Market) 225
  Frankfurt 163
  Glenelg 260
  Great Yarmouth 198, 211, 217, 219, 222
  Hamburg 57, 97, 162, 165, 168
  Hanover 57, 163
  Kilmore 237
  Launceston 237–238
  Leicester 199
  Leipzig 163
  Madras (Chennai) 164, 232
  Manchester 198–199, 208–217
  Manly 245–247
  Melbourne (Bourke Street) 237
  Melbourne 250–257, 260, 263
  Morecambe 198, 214, 217

Index

Munster 163
Napier 2, 232
New York 16
Paris (Boulevard Montmatre) 167
Paris (Jardin d'Acclimatation 162, 165–167
Perth 242
Regent's Park (Fish House) xii, 20, 28–39, 42–58, 61, 63, 146, 160, 167
Rhyl 198, 203, 217
Rome 163
Rothesay 198, 224–225
Scarborough 217–219, 223
Sebastopol 163
Smithsonian Museum 162
Southport 198, 199, 222–224
Surrey Gardens 26, 27, 51
Sutton Coldfield 203
Sydney (Pitt Street) 237
Sydney xi
Tynemouth 203, 223
Vancouver 232, 269
Vienna 163
Westminster 146, 203–207, 217, 220
Zurich 163
Aristotle 2
Arnold, Matthew 115
Arthur, Prince 185
Atkins, Anna 117

Ball, Anne 70, 119
Ball, Mary 70
Ball, Robert 65, 70
Ballantyne, Richard Michael 132
Barnum, Phineas T. 16, 82, 162
Beche, Henry De La 139, 145
Bede, Cuthbert 45
Beneden, Pierre-Joseph van 99, 148
Bentham, Jeremy 5, 6, 271
Bergh, Henry 5

Birch, Eugenius 186, 217
Bisiker, William 126
Bjerring, Nicholas 171, 188
Blacker, William 154, 155
Blunt, Wilfred 32, 54
Boucicault, Dion 177
Bowerbank, James 29
Bradbury, William 117
Braithwaite, Victoria 270
Briggs, Asa 22
Brown, Bill 22
Brown, Culum 270
Brunner, Berndt 11
Buckland, Frank 16, 144, 145, 146, 151, 173, 180, 185, 187, 210, 222, 251, 260
Buckland, William 144

Cameron, Julia Margaret 126
Campbell, Thomasina 120
Cantor, Theodore 232
Carbonnier, Pierre 83
Carlyle, Jane 92
Carlyle, Thomas 92
Catlow, Agnes 122, 133
Clarke, Louise Lane 117
Clealand, James 141
Clough, Arthur Hugh 90
Cockayne, Emily 22
Colley, Anne 22
Coup, William Cameron 16, 177, 207
Covent Garden 80, 93, 226
Crary, Jonathan 23
Croall, Alexander 117
Cruikshank, George 45
Cuming, Hugh 130
Cunningham, Peter 39
Cura, Luigi 84
Cuvier, Georges 13, 153

Damon, William 98
Darwin, Charles 8, 18, 132, 165
Davidson, Richard 126
Dickens, Charles 15, 46
Dickens Jr, Charles 208
Dohrn, Anton 148, 228
Dorner, Herman 178
Dyce, William 113–115

Edwards, Arthur Mead 109
Eliot, George 109, 134, 139
Ellis, Helen 12

Figuier, Louis 165
fish farms
  Concarneau 151
  Crystal Palace 177, 184
  Howietoun 151, 261
  Salmon Ponds, Tasmania 261–262
  South Kensington 144–146, 151
  Twickenham 151
Fitzgibbon, Edward 155
Flack, Andrew 7
Flanders, Judith 101
Ford, Caroline 249
Francis, Eliza Warren 175
Francis, Francis 151, 155, 193, 261
Fraser, Robert William, Rev. 34–35, 43

Garnett, Richard 132
Gates, Barbara 13, 19
Gatty, Margaret 119, 121, 137–140
Gifford, Isabella 119
Gilbert, James 127
Gooday, Graeme 89
Gordon-Cumming, Eliza 116
Gosse, Edmund 31, 142
Gosse, Elizabeth Bowes 13

Gosse, Philip 12, 15, 17, 31–32, 34, 43, 46, 52, 69, 77, 85, 87, 111, 118, 130, 140–142
Gould, John 33, 37
Granata, Silvia 10, 20, 76, 106
Grasset, Eugène 20
Gray, Elizabeth 116
Great Exhibition 25, 37, 168
Greville, Richard 120
Griffiths, Amelia 116, 119, 120, 121
Grootenboer, Hanneke 123
Günther, Albert 126, 232

Haeckel, Ernst 20
Hall, Thomas 85, 93
Hamera, Judith 11
Hamlyn, John 84
Hamon, Philippe 39
Hannaford, Samuel 233
Hannay, James 23
Harper, John 15, 84, 93–94, 136
Harris, Charles 103
Harter, Ursula 11
Harting, James 124
Harvey, William Henry 116, 118–120
Hastie, Jane Alexia 125–126
Hawkins, Benjamin Waterhouse 168
Heath, William 133
Hervey, Alpheus Baker 128
Hibberd, Shirley 86, 94, 106
Hilton, Timothy 114
Hogg, Jabez 88
Hooker, Joseph Dalton 18
Hooker, William 145
Hornell, William 150
Houghton, William 99
Hugo, Victor 173
Hunt, William 207
Hutchins, Ellen 119
Huxley, Thomas 18, 141, 149, 239

Index

Jamrach, Charles 84, 129
Johnston, George 35, 77, 120, 134
Johnstone, William Grosart 117
Jones, Thomas Rymer 17, 98
Jukes, Joseph Beete 132

Keeling, Clinton 53
Kelaart, Edward Frederick 62
Kelson, George Mortimer 155
Kilburn, William 118
King, Amy 43
Kingsley, Charles 73, 82, 118
Kingsley, Mary 126
Krefft, Gerard 243

Lankester, Edwin 149
Lear, Edward 120, 135
Lee, Henry 2, 174, 177, 185, 187, 207
Lee, Sarah Bowdich 13, 126, 152
Lewes, George Henry 134–137, 139, 144
Lloyd, William Alford 28, 32–34, 53, 55, 65, 72, 80, 83, 85–88, 90, 92, 92, 95, 99–101, 102, 106–107, 110, 139, 161, 165–177, 182–183, 188, 196, 200, 204, 220, 228, 239, 260
Loisel, Gustave Antoine Armand 81, 164, 168
Lord, John Keast 176, 185
Lorenzi, Camille 83
Loudon, Jane 122
Lovell, Matilda Sophia 136

Maitland, James 151
marine research stations
　Aberdeen 149
　Cullercoats 149
　Granton 149
　Jersey 63, 150, 220
　Millport 149

Monte Carlo 147
Naples 58, 148–149, 228
Ostend 148
Piel-in-Barrow 149
Plymouth 148–149, 172
Port Erin 149
Puffin Island 149
Roscoff 148
St Andrews 149
Martin, John 156
Mayhew, Henry 84–85, 130, 133
Melville, Herman 25
Meredith, Louisa Anne xii, 263
Merrill, Lynn 20
Millais, John Everett 135, 204
Mitchell, David 27–29, 31, 34, 36–37, 39, 46, 56, 64, 69, 165, 199
Möbius, Karl 97
Mogg, Edward 39
Moine, Fabienne 127
Montizon, Juan, Comte de xii, 35, 51
Moody, Sophy 122

Napoleon III 187, 222
Newland, Henry 155
Nicolson, Adam 7, 12
Nightingale, Florence 90
Norris, Frank 173
North, Marianne 126

Obaysch 27, 36, 44, 47, 91
Ogilby, James Douglas 247
Olsen, Chris 20
Owen, Richard 18, 77, 90, 165, 180, 220

Paterson, Helen 121
Percival, James Gates 132
Philpot sisters (Elizabeth, Mary and Margaret) 116
Playfair, Robert 232

Pliny the Elder 2
Pratt, Anne 118
Pritchard, Zarh 20

Reay, John 44
Reiche, Henry 16, 177
Reiss, Christian 10
Roberts, Mary 118, 123
Rose, Frederick 173

Samuel, Marcus 129
Sandeman, Fraser 155
Saville-Kent, William xiii, 63, 65, 150, 185, 187–189, 196, 204–206, 210–213, 220, 239–242, 261
Schomburgk, Richard 242
Sea Life centres 268
Sea World 268
Seder, Anton 20
Senior, Emily 35
Senior, William 262
Sheppard, Harriet Campbell 119
Sherson, Errol 206
Sherwin, James 209
Shetterly, Susan 12
Sinel, Joseph 150
Singer, Peter 3–7, 269, 271
Small, H. Beaumont 82
Smiles, Samuel 32
Smith, Jonathan 50
Sneddon, Lynne 270
Souef, Albert le 253
Southwick, Emma 186

Sowerby, George 124, 129, 130, 162–163
Spencer, Herbert 133
Stoddart, Thomas 155
Stimpson, William 78
Stott, Rebecca 12
Sullivan, Arthur 204

Taylor, John Ellor 66, 134, 223, 256
Teltscher, Kate 121
Thompson, John Vaughan 146
Thomson, Charles Wyville 147
Thynne, Anna 45, 77, 100, 111, 132
Trentmann, Frank 22
Turner, Dawson 119
Twining, Elizabeth 122

Vennen, Mareike 11
Victoria, Queen xiii, 50, 168, 170
Villepreux-Power, Jeanne 77

Ward, Nathaniel 73, 78
Warrington, Robert 28, 34, 35, 46, 77, 88
Watney, Helen 86, 140, 142
Webb, Captain 219, 227
Wheatley, Hewett 155
Wilberforce, Samuel 18
Wilde, Oscar 221
Willughby, Francis 44
Wilkinson, Anne 95
Wyatt, Mary 116

Youl, James 261

www.ingramcontent.com/pod-product-compliance
Lightning Source LLC
Chambersburg PA
CBHW031314160426
43196CB00007B/521